The Institute of Biology's
Studies in Biology no. 84

Lysosomes

R. T. Dean
M.A., Ph.D.
M.R.C. Clinical Research Centre,
Harrow, Middlesex

Edward Arnold

© Roger T. Dean

First published 1977
by Edward Arnold (Publishers) Ltd
25 Hill Street, London W1X 8LL

Board edition ISBN 0 7131 2662 0
Paper edition ISBN 0 7131 2663 9

All Rights Reserved. No part of this publication
may be reproduced, stored in a retrieval system,
or transmitted in any form or by any means, electronic,
mechanical, photocopying, recording or otherwise, without
the prior permission of Edward Arnold (Publishers) Limited.

Printed in Great Britain by
The Camelot Press Ltd, Southampton

General Preface to the Series

It is no longer possible for one textbook to cover the whole field of Biology and to remain sufficiently up to date. At the same time teachers and students at school, college or university need to keep abreast of recent trends and know where the most significant developments are taking place.

To meet the need for this progressive approach the Institute of Biology has for some years sponsored this series of booklets dealing with subjects specially selected by a panel of editors. The enthusiastic acceptance of the series by teachers and students at school, college and university shows the usefulness of the books in providing a clear and up-to-date coverage of topics, particularly in areas of research and changing views.

Among features of the series are the attention given to methods, the inclusion of a selected list of books for further reading and, wherever possible, suggestions for practical work.

Readers' comments will be welcomed by the author or the Education Officer of the Institute.

1977   The Institute of Biology,
                                               41 Queen's Gate,
                                               London, SW7 5HU

Preface

Lysosomes are membrane-limited sacs present in most cells of animals and plants, which contain a diverse array of degradative enzymes. As was anticipated more than ten years ago by de Duve, the discoverer of the lysosome, they are important in the destructive phase of the continuous breakdown and resynthesis which most cellular material (such as proteins) undergoes. They are likewise concerned with the destruction of foreign material entering cells, and this constitutes a major element of the defence system of animals, for instance, against invading organisms. And in unicellular organisms, such as Protozoa, they serve as a main digestive system. More recently a wide range of less predictable functions of lysosomes (controlled incomplete degradative processes, intracellular transport and sequestration, and extracellular secretion) has become apparent. And because of the destructive capacity of lysosomal enzymes it is easy to appreciate that malfunction of the lysosomal system can have serious consequences.

This book attempts to provide a clear conceptual outline of the structure and function of lysosomes, and to indicate briefly their involvement in pathological disturbances: Chapters 1–5 describe the natural history of lysosomes, while Chapter 6 considers their functions, and mentions several examples of lysosomal pathology.

London, 1977   R. T. D.

Contents

	General Preface to Series	iii
	Preface	iii
1	The Structure and Distribution of Lysosomes 1.1 The discovery of lysosomes 1.2 Isolation 1.3 Morphology 1.4 Composition 1.5 Organization 1.6 Occurrence	1
2	Physico-chemical Properties of Lysosomes 2.1 Stability and permeability 2.2 The intralysosomal pH	18
3	Formation and Fate of Lysosomes 3.1 Synthesis of lysosomal components 3.2 Translocation of lysosomal components and formation of primary lysosomes 3.3 Transformation of lysosomes 3.4 Cytoplasmic liberation of lysosomal contents 3.5 Degradation of lysosomal constituents	23
4	Endocytosis and Exocytosis 4.1 Phagocytosis 4.2 Pinocytosis 4.3 Exocytosis	32
5	Autophagy and the Accumulation of Materials 5.1 Autophagy 5.2 Accumulation	40
6	Functions of Lysosomes 6.1 Degradation of endocytosed materials 6.2 Degradation of intracellular materials 6.3 Effects of accumulation of materials by lysosomes 6.4 Extracellular activities of lysosomal components	43
	Glossary of Lysosomal Terminology	53
	Further Reading	54

1 The Structure and Distribution of Lysosomes

1.1 The discovery of lysosomes

The biochemist tends to be interested in enzymes, proteins which function as biological catalysts. Usually one of his main problems is that his chosen enzyme is unstable, and rapidly ceases to be capable of catalysing its reaction when he tries to purify it, or even to study it in a tissue homogenate. Christian de Duve was faced with quite the opposite problem in the early nineteen fifties: when he let a homogenate of rat liver stand, his enzyme increased in apparent activity! The enzyme was acid phosphatase, which hydrolyses several organic phosphates, and works best under acid conditions:

$$^{-}O-\underset{\underset{O}{\|}}{P}-O-R + H_2O \rightleftharpoons HOR + {}^{-}O-\underset{\underset{O}{\|}}{P}-OH$$

Undeterred, de Duve showed that this slow activation could be produced much more rapidly by various physical tricks, such as freezing and thawing or adding detergents, and that it was accompanied by a transfer of the enzyme from a state in which it could be centrifuged into a precipitate ('sedimented') at about 20 000 g in 15 minutes, to one in which it was not even sedimented by centrifuging at 100 000 g for 1 hour. de Duve surmised that acid phosphatase was initially contained within a continuous membrane in a sedimentable particle. The enzyme activity in a homogenate is normally very low because the enzyme inside the particle cannot reach the substrate (organic phosphate) which is added to the solution outside the particle (the enzyme is said to be 'latent') but as the membrane breaks progressively on standing, or because of freezing and thawing, the enzyme gains access to the substrate, and activity is increased. Thus the latency is 'structure-linked', in that it depends on the intactness of the lysosomal membrane (Fig. 1–1). Having shown that the sedimentation characteristics of the particle were distinct from those of other known cellular entities (see section 1.2), de Duve proposed that they be called 'lysosomes', and that lysosomes are the sole location of acid hydrolases.

The term was logically chosen: lysosomal enzymes break up their substrates by means of water, i.e. hydro*lyse* them, and are contained

INTACT LYSOSOME   **DAMAGED LYSOSOME**

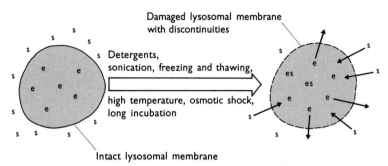

Intact lysosomal membrane

Fig. 1–1 Latency of lysosomal enzymes. In normal lysosomes the intact membrane prevents both access of substrates (s)—e.g. phosphate esters, nucleic acids, proteins, glycosides—to the interior of the lysosomes, and the exit of lysosomal enzymes (e) to the exterior of the lysosomes. Thus enzyme activity is hardly detectable until the membrane has been disrupted by one of the means indicated above.

within particles ('*somes*' from the Greek *soma*). Because of the potentially dangerous nature of lysosomal hydrolases, it is clearly desirable to compartment them, so that they do not have uncontrolled access to cellular materials. Indeed, when lysosomes within cells are broken specifically by experimental manipulation cell death ensues rapidly. However, lysosomes are far from being the 'suicide bags' as which they have been popularized in some quarters: they are 'behind bars' (de Duve), but in no pejorative sense! de Duve was awarded the Nobel Prize in 1974 in recognition of his discovery of lysosomes, and his subsequent work on these and other cellular structures. He gives a delightful description of the early development of the field in the first of the series of books on lysosomes edited by DINGLE and associates (see Further Reading).

1.2 Isolation

Although de Duve was able by 1960 to show that lysosomes were distinct particles, using centrifugation techniques, and that several hydrolases most active under acid conditions besides acid phosphatase were lysosomal, he did not obtain completely pure lysosomes.

The general problems of isolating cellular organelles in an undamaged state, have been discussed in previous volumes (nos. 9 and 31). Some of the difficulties are: how to break cells without breaking the organelle of interest, how to maintain the organelles intact once they are removed from their normal cellular milieu, and how to separate the organelles from each other. By now we have a fair idea of how best to deal with the first two points, though no method could be described as ideal. As far as lysosomes are concerned, the problem of how to completely separate the organelle from others remains serious.

Centrifugal techniques depend for success primarily on differences in size and/or density of the particles to be separated. Unfortunately lysosomes are similar in both these respects to mitochondria, of which there are rather more in most cells. The differential centrifugation scheme used by de Duve in 1955 in characterizing lysosomes is shown in Fig. 1–2, which indicates the similar behaviour of mitochondria and lysosomes.

However, certain functional characteristics of lysosomes can be used to modify their behaviour in centrifugation, and obtain their complete separation from mitochondria (and other structures). For instance, as discussed in Chapter 4, some substances can enter cells by the process of phagocytosis, in which they become encapsulated in a membrane-bounded vesicle which fuses with lysosomes. If the substance has a very low or very high density, its accumulation in lysosomes will alter their density sufficiently for it to be distinct from that of mitochondria. Thus if particles are centrifuged to equilibrium on density-gradients (usually of sucrose) mitochondria and modified lysosomes can be separated, and pure lysosomes obtained (Figs. 1–3 and 1–4). Alternatively, there are now means of altering the density of mitochondria to the same end. Lysosomes show such physical and chemical diversity (e.g. Figs. 1–6 and 1–7) that any purified lysosomal fraction is unlikely to be fully representative of the lysosomal system. Some of the enzymatic criteria which, in conjunction with morphological examination, are used to assess the purity of such preparations, are illustrated in Fig. 1–5, which describes a particular preparation of rat liver lysosomes.

It is now clear that most acid hydrolases are almost entirely restricted to lysosomes, as de Duve anticipated. β-Glucuronidase is a notable exception, showing substantial activity in the endoplasmic reticulum of most cell types. Other lysosomal enzymes do show low, but detectable activity in endoplasmic reticulum (section 5), but this may be merely due to enzyme in transit to lysosomes after synthesis on rough endoplasmic reticulum (see section 3.1).

1.3 Morphology

In electron micrographs of tissue sections there can be seen numerous heterogeneous cytoplasmic vacuoles with single membranes, which are clearly distinct from mitochondria, and from those small vacuoles with crystalline inclusions (peroxisomes). Many are lysosomes, but they can only be conclusively identified as such by means of enzyme staining (see Figs. 1–6 and 1–7 for examples of such sections).

Acid phosphatase was originally most often used, but now the activity of several other acid hydrolases can be demonstrated at both the light and electron-microscope levels. For several enzymes, it is difficult to obtain

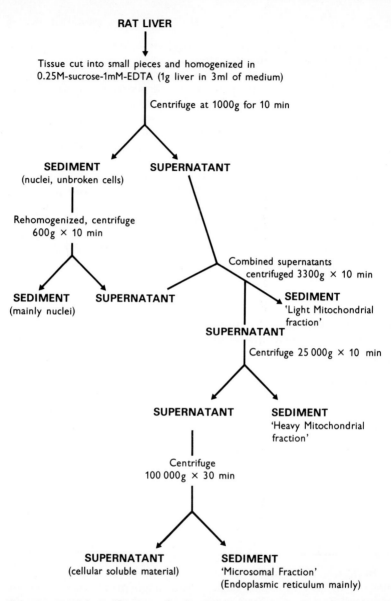

Fig. 1-2 Partial purification of subcellular components from rat liver by differential centrifugation. 0.25 M-Sucrose provides osmotic protection (see text), while EDTA (ethylene diamine tetra-acetic acid) as its sodium salt, sequesters trace metals which inhibit or destabilize some enzymes. All operations are conducted at 4°C. As the names 'Light' and 'Heavy Mitochondrial Fraction' imply, two fractions are obtained which consist largely of mitochondria. The first of these contains a higher proportion of lysosomes, and constitutes a partially purified lysosomal fraction. Such a fraction is often used for further purification.

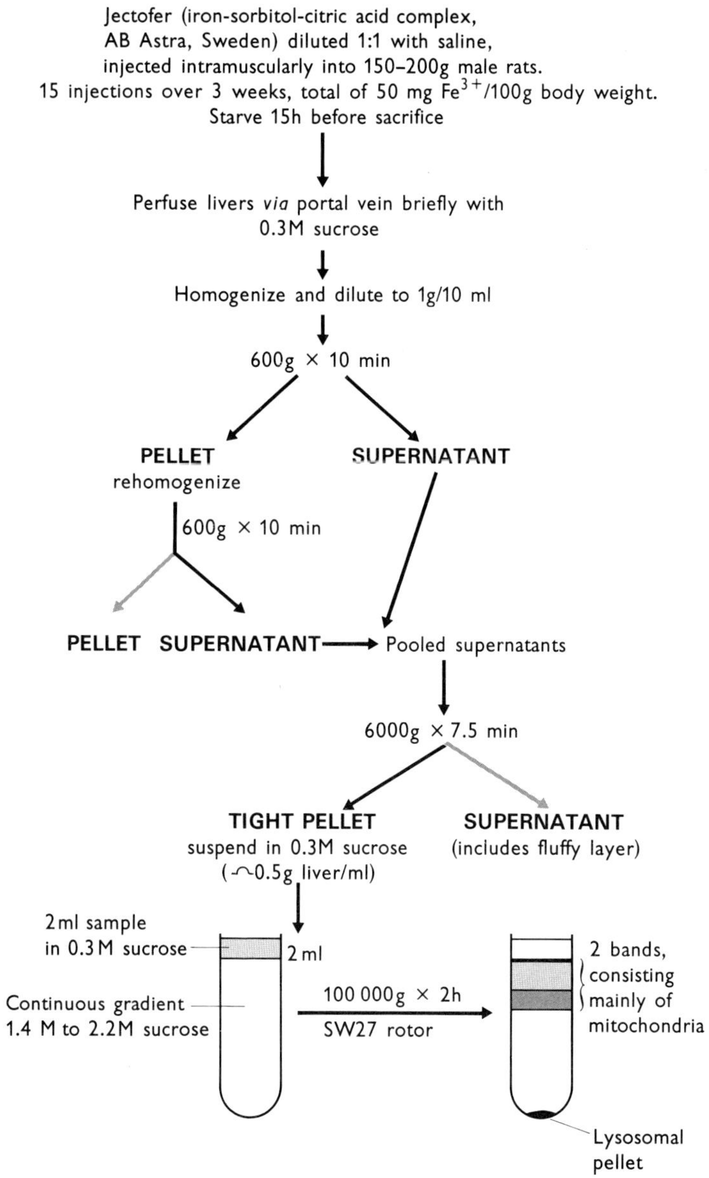

Jectofer (iron-sorbitol-citric acid complex,
AB Astra, Sweden) diluted 1:1 with saline,
injected intramuscularly into 150–200g male rats.
15 injections over 3 weeks, total of 50 mg Fe^{3+}/100g body weight.
Starve 15h before sacrifice

↓

Perfuse livers *via* portal vein briefly with
0.3M sucrose

↓

Homogenize and dilute to 1g/10 ml

↓

600g × 10 min

↙ ↘

PELLET **SUPERNATANT**
rehomogenize

600g × 10 min

↙ ↘

PELLET **SUPERNATANT** ⟶ Pooled supernatants

↓

6000g × 7.5 min

↙ ↘

TIGHT PELLET **SUPERNATANT**
suspend in 0.3M sucrose (includes fluffy layer)
(∼0.5g liver/ml)

2 ml sample in 0.3M sucrose — 2 ml

Continuous gradient 1.4 M to 2.2M sucrose

100 000g × 2h
SW27 rotor

2 bands, consisting mainly of mitochondria

Lysosomal pellet

Fig. 1–3 Isolation of rat liver lysosomes loaded with colloidal iron (after ARBORGH *et al.* (1973). *FEBS Lett.*, **32**, 190–194).

6 MORPHOLOGY

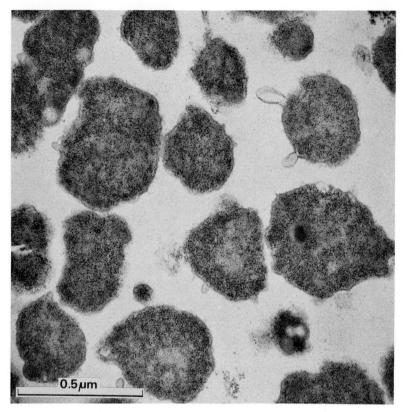

0.5 μm

Fig. 1-4 Purified iron-loaded lysosomes. The lysosomes show diverse shapes and highly irregular membrane contours. No contaminating organelles, such as mitochondria, are apparent. Electron micrograph of Dr H. Glaumann, reproduced with his kind permission.

specific substrates with which to demonstrate the localization of the enzyme unambiguously, because other enzymes also transform the substrates. Therefore 'immunological' techniques for localization of lysosomal enzymes are becoming increasingly important. These techniques rely upon the fact that when foreign molecules are injected into higher animals, such as man, the animal synthesizes molecules (called antibodies) which circulate in the blood, and are capable of reacting specifically with the injected molecule and not with other molecules. They constitute a normal defence mechanism of animals against foreign materials and organisms (such as bacteria). This so-called 'immune response' is exploited medically to give protection against diseases (by means of vaccinations) and is also exploited scientifically. If a

Fig. 1–5 Characteristics of purified iron loaded lysosomes from rat liver (data from ARBORGH *et al.* (1973) *FEBS Lett.*, **32**, 190–194).

Component (organelle represented)	Recovery in whole procedure %	Yield in lysosomal fraction %	Relative specific activity in the lysosomal fraction	Likely contributions to protein of lysosomal fractions %
Protein	100.5	.34	1	
Acid phosphatase (lysosomes)	103.0	11	33	
Aryl sulphatase (lysosomes)	112.0	12	33	
Cathepsin D (lysosomes)	107.0	13	35	
NADPH-cytochrome C reductase (microsomes)	N.G.	.007	.022	0.5
Succinate-Cytochrome C reductase (mitochondria)	N.G.	.033	.091	1.8
D-Amino acid oxidase (peroxisomes)	N.G.	.23	.615	1.5
Summed contaminants				≥ 3.8

Important biochemical criteria of purifications are as follows: the summed recoveries of enzymes and protein in the whole procedure should be determined, and should be between 90–110% if interpretation is to be meaningful. Enzyme recoveries of greater than 100% often indicate the separation of enzyme from inhibitors; on the other hand, a low enzyme recovery usually indicates an unstable enzyme. Obviously the absence of an unstable mitochondrial enzyme from a lysosomal fraction is not satisfactory evidence for the absence of mitochondria. However, when stable enzymes which are characteristic of known organelles are shown to be absent, and yet recovered elsewhere, it can be claimed that the lysosomal preparation is free of contamination by the organelle concerned. Such information is usually presented as the enzyme relative specific activity (RSA) in the fractions: specific activity (SA) is activity per unit of protein, while RSA is the ratio of SA of the enzyme in the fraction to its SA in the homogenate. Thus in a lysosomal preparation, one requires lysosomal enzymes to show high RSA values (three are shown above) and enzymes from other organelles to show low RSA values (three are shown above). Such data can also be used to estimate the amount of contaminating protein each organelle could contribute. In this table contaminants seem only to account for about 4% of the protein of the lysosomal fraction, but there are two shortcomings to the data. Firstly, no 'marker' enzyme for the cell membrane was assessed, and so the degree of contamination by plasma membrane is unknown. Secondly the total recoveries of the three marker enzymes for non-lysosomal structures were not given (N.G.).

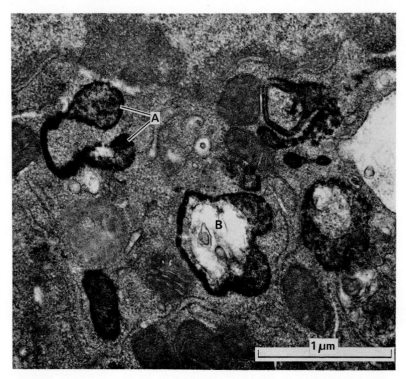

Fig. 1–6 Autophagy and the morphological heterogeneity of lysosomes. This section of a histiocyte from the subcutaneous region of a mouse injected with peroxidase (given subcutaneously) has been stained for acid phosphatase. A lysosome apparently 'wrapping' round some cytoplasm (A), contains enzyme activity through the matrix within the single membrane. The acid phosphatase of a second lysosome (B), perhaps the product of such a wrapping process, is found only between its pair of membranes, the inner of which will eventually be degraded, allowing the lysosomal enzymes to attack the entrapped substrates. Several other lysosomes are visible. By kind permission of Dr K. Ogawa.

single material is injected, then the antibodies will react only with that material. Thus antibodies specific for lysosomal enzymes can be obtained (with considerable effort!) and if coloured or electron absorbing materials are chemically linked to the antibodies, the combination can be used to localize the lysosomal enzyme either by light microscopy (with a coloured material) or electron microscopy (with an electron absorbing material). Fig. 1–8 shows the localization of the lysosomal proteinase (see section 1.4) cathepsin D in rabbit fibroblasts, demonstrated by light-microscopy using a sensitive modification of this method.

Other characteristics often used to identify lysosomes, for instance

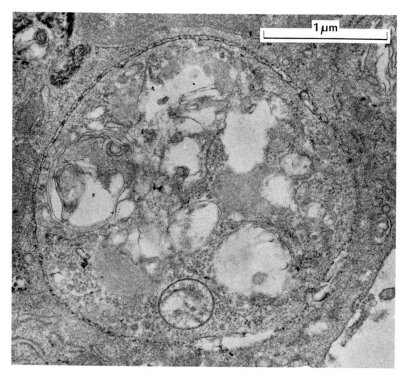

Fig. 1–7 An autophagic vacuole with two membranes and acid phosphatase only between the membranes. A wide range of cellular membrane fragments, about to undergo digestion, are present within the lysosomal membranes. From a section of a mouse liver histiocyte, animal treated as for Fig. 1–6, and the plate kindly provided by Dr K. Ogawa.

staining by heavy metal ions (for electron microscopy) and with fluorescent dyes such as acridine orange (for fluorescence microscopy), are not always satisfactory as other organelles may stain.

Although lysosomes vary greatly in size and shape (Figs. 1–6 and 1–7) there are some common morphological features: the membrane (and that of organelles with which lysosomes interact by fusion (section 3.4) such as plasma membranes) is thicker than that of other organelles, and immediately within it there is usually an electron-lucent area. Several morphological variants are described in later sections; the transformations, following phagocytosis, for instance, can be used to introduce identifiable materials into lysosomes (thorotrast (thorium dioxide) or colloidal gold particles, or the enzyme horse-radish peroxidase).

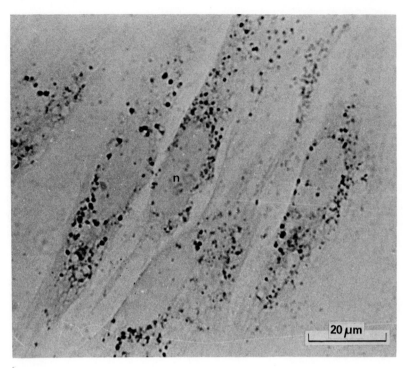

Fig. 1–8 Localization of cathepsin D in rabbit fibroblasts. The sites of the enzyme were revealed by means of a double-antibody technique, in which the fixed cells were incubated first with a sheep antibody preparation from a specific antiserum to rabbit cathepsin D, and secondly with a pig antibody conjugated with horseradish peroxidase, in which the antibody was obtained from an antiserum to sheep IgG. After washing, the cells were reacted for peroxidase and examined by bright-field microscopy. Particulate staining is observed, indicative of a lysosomal distribution of cathepsin D. n, nucleus. Photograph kindly provided by Dr A. R. Poole.

1.4 Composition

Although it is clear that the lysosomes of some cell types contain some specialized enzymes, most animal cells possess a common but diverse array of lysosomal enzymes, almost all of which are hydrolases. About 60 enzymes are known to be present in the lysosomes of one or more cell types: there are several proteinases, glycosidases, nucleases, phospholipases, phosphatases and sulphatases. These classes of lysosomal hydrolases are characterized in Fig. 1–9. The pH optima of the enzyme activities are normally in the acid range, although some enzymes active at neutral pH are known.

§ 1.4   COMPOSITION 11

In general the enzyme armoury is sufficient to degrade most proteins to small peptides and amino acids, carbohydrate moieties (in glycoproteins (proteins with attached sugar chains), proteoglycans (materials in bone and cartilage with large amounts of polymeric carbohydrate attached to protein cores) and glycolipids (complex lipids containing sugars)) to monosaccharides, nucleic acids to nucleosides and phosphates, complex lipids to free fatty acids, phosphates, etc., and to remove phosphate groups from many other materials including phosphoproteins.

Class	Enzyme	Typical substrate	Reaction	Optimal pH
Oxidoreductases acting on hydrogen peroxide as acceptor	Peroxidase	Benzidine, proteins as donor: H_2O_2 as acceptor	Donor oxidized; H_2O_2 converted to H_2O	5.5
Hydrolases acting on ester bonds (esterases)	Acid phosphatase (several types)	p-nitrophenyl phosphate	Liberates phosphoric acid from a variety of organic phosphates	3.0–6.0
	Cholesterol esterase	Cholesterol oleate	Liberates fatty acid and cholesterol	4.0
	Deoxyribonuclease II	DNA	DNA split to 3'-phospholoigonucleotides	4.5–5.5
Hydrolases acting on glycosyl bonds (glycosidases)	β-glucuronidase	β-D-glucuronic acid-phenolphthalein	Liberates terminal residues of β-D-glucuronate	4.0–5.5
Hydrolases acting on peptide bonds (peptidases and proteinases)	Cathepsin D	Proteins	Liberates small peptides	3.0–3.5

Fig. 1–9 The main classes of lysosomal enzymes. Only a few examples amongst the sixty or so known lysosomal enzymes are given; they are not all necessarily present in the lysosomes of every cell: for instance, peroxidase is mainly present in leucocytes, but little elsewhere.

Most lysosomal enzymes are negatively charged glycoproteins: the significance of this is still unclear, although two possibilities are discussed in subsequent sections. Variation of constitution of the carbohydrate moiety of lysosomal enzymes is an important source of the immense array of multiple forms most show (differing slightly in charge but often not in enzymatic characteristics). Fig. 1–10 gives some quantitative data on the enzymatic constitution of liver lysosomes. But it should be noted that lysosomes even in single cell types are quite variable in their enzymatic constitution: for instance, proteinase-rich and glycosidase-rich lysosomes are discernible in the protistan *Tetrahymena* and less clear-cut heterogeneity is widespread. When lysosomes of a tissue composed of several cell types (e.g. liver) are considered, the diversity is still more pronounced.

Enzyme		Enzyme protein as % of lysosomal protein	Enzyme concentration in the intralysosomal space mM
Proteinases	Cathepsin D	10.6	0.78
	Cathepsin B	11.8	1.54
Glycosidases	β-Glucuronidase	3.4	.04
	N-Acetylhexo-saminidase (all forms)	1.5	.04
	α-Fucosidase	.1	.002

Fig. 1–10 Estimates of the quantitative contribution of certain enzymes to lysosomal protein in human liver. The enzyme concentration values are interesting for two reasons: firstly that the various enzymes show widely differing concentrations, secondly that the values are remarkably high, because of the very small volume within which the enzymes are trapped. Such high concentrations allow rapid action, and facilitate concerted action of enzymes; they are also extremely difficult to reach experimentally, and illustrate one of the many important differences between *in vitro* and *in vivo* conditions.

The enzymes are probably mainly present in the soluble phase in the interior of lysosomes (see section 1.5) and several other materials are present there also. For instance, lipoproteins constitute a substantial part of the protein of isolated rat kidney and liver lysosomes. Several different lipoprotein fractions, mostly polyanionic (with many negative charges), are found. The denser lipoproteins are probably formed from the lighter by loss of lipid. Substantial amounts of dialysable 'phosphatido-peptides' (containing phospholipids and amino acids) have also been found.

Several cationic (positively charged) proteins are present in the azurophil (lysosomal) granules of polymorphonuclear leucocytes, one of the white blood cell types; these include the glycosidase lysozyme, and the proteinases elastase and cathepsin G, in man. Some of the cationic proteins are bactericidal and some bactericidal proteins (not all of which are yet known to be enzymes) are also present in lysosomes of platelets, the very small blood cells concerned with clotting. It is notable that these bactericidal proteins are restricted to the blood cells which form the first line of defence against invading bacteria, etc. They are not found in lysosomes of liver or most other tissues. The membrane of the lysosome has qualitatively a fairly usual composition (see no. 27 in this series) and thus contributes several components. Amongst these the phospholipid and cholesterol contents of lysosomes have been analysed: choline phospholipids and sphingomyelin predominate, and the phospholipid to protein weight ratio is of the order of 0.3. The presence of cholesterol and sphingomyelin is notable, because these materials are virtually absent from other cytomembranes, with the exception of plasma membranes. This observation is consistent with the known exchanges of membrane between lysosomes and plasma membranes occurring during endocytosis and exocytosis (see section 3.4).

Ions of several metals (iron, manganese, etc.) are found normally in lysosomes, at concentrations higher than in other organelles. In some cases (e.g. Wilson's disease, in which copper is stored in the liver) the accumulation of a metal may be a consequence of some pathological lesion. But normally the lysosomes may be protecting the cell from the deleterious effects of such ions, by sequestering them.

Several materials occurring within lysosomes as a result of the activities of the lysosomal system can be identified. For instance, cytochromes from the mitochondrial electron transport chain, undergoing digestion (section 6.1), can be detected by their spectral characteristics; and ferritin, a cytoplasmic protein which is quite resistant to lysosomal enzymes, is often observed. Lipofuscin, a complex compound containing lipids and iron (originally termed 'ageing pigment'), which is apparently indigestible, accumulates in some lysosomes in old organisms. A lysosomal amino acid pool has been reported, and recently also a pool of protein-degradation products (labelled *in vivo*), presumably oligopeptides which do not rapidly pass through the lysosomal membrane.

1.5 Organization

The intralysosomal distribution of materials can be investigated by breaking lysosomes, preferably by hypotonic treatment, and separating sedimentable and non-sedimentable material by centrifugation (100 000 g × 60 min normally).

The membrane. The membrane material, with some gains and some losses, is represented by the sedimentable fraction. Roughly 35% of lysosomal protein, 75% of phospholipids, and 70% of cholesterol are recovered in the 'membrane' fraction. Both lysosomal and plasma membranes contain an excess of acidic over basic amino acids, but detailed analysis of complex lipids emphasizes the dissimilarity between the composition of lysosomal and plasma membranes. Thus lysosomal and plasma membranes have many common features, but are quite distinct.

There is substantial carbohydrate material in the membrane, including about 16 µg sialic acid/mg protein, which is rather more than in other cellular membranes. Lectins, which are plant proteins capable of binding to particular sugars in sugar chains of glycoproteins and other sugar complexes, have been exploited experimentally in studying the distribution of the carbohydrate within the lysosomal membrane. When binding of such lectins to intact lysosomes is compared quantitatively with binding to broken lysosomes, it is apparent that most binding sites are on the internal face of the lysosomal membrane. This is consistent with the results of staining for carbohydrate with colloidal iron. There is some binding of lectins to the outside surface however, and the mobility characteristics of intact lysosomes in electrical fields support this. During such particle electrophoresis in aqueous media, only the accessible (i.e. surface) charges of the particle affect mobility. Lysosomes have a nett negative surface charge, as judged by the technique, and this charge is reduced by digestion with neuraminidase (which removes negatively charged sialic acid residues). As the neuraminidase neither breaks nor enters the lysosomes, the results indicate that some sialic acid is on the external surface of lysosomes.

The enzymes. There is fragmentary information available on the intralysosomal distribution of enzymes. Because of the ambiguities of the fractionation procedures, the fact that a third of the 'membrane' protein can be removed by simply washing with 0.9% NaCl ('physiological saline': materials removed by this are unlikely to remain attached *in vivo*!) and the demonstration of reversible pH and ionic strength dependent binding of β-glucuronidase and acid phosphatase to lysosomal membranes (mainly dependent on membrane sialic acid), this information cannot be considered entirely satisfactory. Similarly, qualitative electron microscope histochemistry is inadequate, but quantitative histochemistry may soon allow rigorous study. Certain enzymes do seem to be primarily membrane bound. Some, including an esterase, may be on the outside of intact lysosomes. This disposition is the simplest explanation of the fact that the enzyme activities are fully expressed in intact lysosomes, while other enzymes in the same preparations show the normal 'structure-linked latency' (section 1.1).

A possible function of the carbohydrate moiety present on most lysosomal enzymes is to control the degree to which an enzyme interacts with the inside of lysosomal membranes. Such interactions depend largely on two factors: the charges on the membrane region in question and on the enzyme (which should be complementary), and the availability of immense areas of hydrophobic material in the membrane which can allow the incorporation of hydrophobic protein surfaces. Extensive glycosylation of a lysosomal enzyme should reduce its affinity for the hydrophobic membrane regions, and extensive sialylation (introducing more negative charges on the enzyme surface) should reduce electrostatic interaction with the negatively charged internal membrane surface. In view of the observation that the carbohydrate moieties of total lysosomal proteins are degraded more rapidly than the proteins, it is possible that the degree of binding of individual proteins might be changed by glycosidase action.

Some evidence consistent with this is available in the case of acid phosphatase: differential release into the non-sedimentable phase, by methods of Fig. 1–1, of distinct multiple forms has been observed from lysosomes of macrophages and of liver. This indicates differential binding of the forms. Similarly, lysosomal enzymes in kidney fractions are found to be progressively more easily released (by freezing and thawing ten times) in the order: rough endoplasmic reticulum < smooth endoplasmic reticulum < Golgi < lysosomes, with the implication of different partitioning between membrane and sol in the fractions. Further, the soluble enzymes consist mainly of anionic multiple forms, while the bound hydrolases are more cationic, implying a role of electrostatic interactions in membrane binding. The changed partition in the different fractions may be due to progressive glycosylation (producing increased electronegativity, as sialic acid is incorporated) during transport, though this has not received rigorous study.

Evidence for a discrete protein component of microsomes by which β-glucuronidase may be bound to the membrane has been obtained. This 'binding protein' is not present in lysosomes, and this implies a quite distinct possibility for controlling intra-lysosomal distribution. Genetic control can operate directly and specifically on such binding, while this is not possible with the glycosylation-deglycosylation mechanism discussed above.

Effects of membrane binding on enzymes. It is known that the binding of enzymes to charged solid phase supports (such as membranes) can have radical effects on the characteristics of the enzyme activity. For instance, the dependence of the activity on the acidity of the bulk fluid phase in which the membrane is situated, will be changed, partly because a local environment of quite distinct acidity is produced in and adjacent to the membrane where the enzyme is present, but also because the enzyme may itself be modified in shape or function. Other changes in the kinetics and specificity of the enzyme may ensue also. These may be of considerable physiological importance. Several examples of such intralysosomal effects are known already. For instance, the lysosomal proteolytic activity which can interconvert two forms of the cytoplasmic enzyme fructose-1,6-diphosphatase acts optimally at pH 6.5 when membrane bound, but at pH 4.5 when released into the soluble phase.

Another possibility allowed by the occurrence of membrane-bound

16 ORGANIZATION § 1.5

enzymes is that of the control of disposition of enzymes relative to their substrates and to each other. Thus intact brain lysosomes are capable of multistep hydrolysis of certain complex lipids; this capacity is destroyed by ultrasound treatment (sonication). Sonication not only breaks lysosomal membranes but also releases some materials from the membrane, and this latter process seems to be responsible for the loss of hydrolytic capacity, indicating that either controlled enzyme disposition relative to other enzymes or to the substrate (or both) is required for hydrolysis. Some enzymes of this multistep process may actually *depend* on lipid for activity and/or stability, as do many enzymes. A clear example of this dependence is the lysosomal enzyme hydrolysing the complex lipid

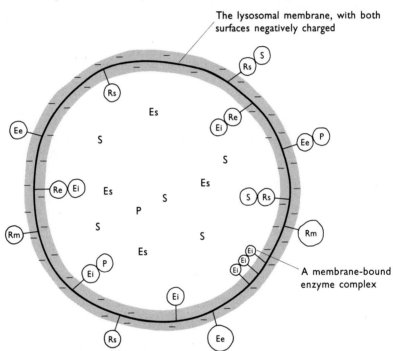

Fig. 1–11 The functional organization of lysosomes. Enzymes may be present either on the external membrane surface (Ee), the internal surface (Ei), or in the internal soluable phase (Es). Their binding to the membrane may in some cases be mediated by specific receptors (Re). Amongst membrane bound enzymes will probably fall some ATP-ase-ion transport enzymes, and some other carrier proteins. Substrates (S) may also be soluble or membrane bound. Substrate binding sites (Rs) may be either specific or non-specific; in contrast, membrane receptors (Rm) for modulators of function (hormones, cyclic nucleotides, etc.) are likely to be quite specific. Soluble proteins (P) which modify the activity of both Ei and Ee are anticipated.

glucocerebroside. This enzyme is constituted from a membrane-bound catalytic protein (C) and a soluble, heat-stable glycoprotein (P) which controls the activity of C. Although the mechanism of interaction of P and C is not known, it is likely that this mechanism of control of enzyme activity, involving an interacting protein molecule, together with the others outlined above, will prove to be of considerable importance.

Summary. Fig. 1–11 indicates some of the likely structural organization of lysosomes.

1.6 Occurrence

Lysosomes are known to occur in most animal phyla, and will probably be found in the remainder. In laboratory animals such as the rat, they are present in almost every cell type which has been studied carefully. In some they take on quite specialized forms, as in skeletal muscle, where most lysosomal enzyme activity seems to be present in localized regions of the modified endoplasmic reticulum (the sarcoplasmic reticulum) of these cells.

The large vacuole of many plant cell types is a modified lysosome, though it has other functions which are not shared by lysosomes of animal cells, mainly to do with nutrient and fluid storage. Unicellular Protista have lysosomes, very similar to those of higher animal cells, which function as an intracellular digestive tract. The alimentary canal of most animals relies very little on lysosomal enzymes in digesting food. Only prokaryotes, such as bacteria, lack lysosomes: but even in bacteria hydrolases are often concentrated (for instance near the cell membrane) in such a way as to imply some relationship with lysosomal packaging.

The attributes of lysosomes discussed in subsequent chapters are those observed in studies of higher animal cells, Protista, and lower animals. Most, however, are probably shared also by lysosomes of plants.

2 Physico-chemical Properties of Lysosomes

2.1 Stability and permeability

As mentioned already, the latency of lysosomal enzymes in intact lysosomes reflects the inability of the substrates used to penetrate lysosomal membranes. In order to maintain such latency in a suspension of lysosomes, it is necessary not only to equalize osmotic pressures inside and outside lysosomes (which can be achieved by most agents at suitable concentration) but also to ensure that the compound providing the external osmotic pressure does not penetrate lysosomal membranes. For, if such penetration occurs, lysosomal swelling by water influx will accompany the entry of the external compound, and breakage of the lysosomes ensues. It has been shown that increases in activity during incubation in suspension do reflect such progressive breakage, and not changes in the permeability of lysosomal membranes.

As a corollary of these observations, the ability of a compound to provide osmotic 'protection' at isosmotic concentration, is an index of its lack of permeation into lysosomes: the better the protection, the lower the permeability. Such experiments indicate that 0.25 M sucrose is a good protectant; it is widely used in the preparation of lysosomes (Fig. 1–2), and in experiments on lysosomes.

In addition, it seems that disaccharides or larger oligosaccharides and polysaccharides, tripeptides or larger oligopeptides, and proteins, and charged anions, penetrate very little, and thus can provide good osmotic protection. In general, compounds smaller than about 300 daltons can penetrate (sugars, amino acids, etc.), and those larger than this cannot. Monovalent cations (Na^+, K^+) penetrate quite freely, but because anions (Cl^-) seem not to penetrate well, there is little *nett* entry of KCl or NaCl (equivalent fluxes occurring both inwards and outwards). Thus even KCl and NaCl can provide osmotic protection.

There is selectivity in cation permeability: $H^+ >> K^+ > Na^+$ is the sequence of magnitudes in permeability at 37°C, neutral pH. Permeabilities of lysosomal membranes are changed at lower temperature: for instance K^+ permeability is increased substantially. Thus experiments performed at lower temperatures give results which cannot always be applied to the situation normally faced by lysosomes in higher animals which are homoiothermic, operating at temperatures such as 37°C! The biochemist usually prefers to operate at low temperatures (as in Figs. 1–1 and 1–2) in order to reduce the rates of destruction and breakage of components (which usually show high Q_{10} values): but he has continually to bear in mind that things may be different at higher temperatures.

The question arises whether these permeability characteristics of isolated lysosomes studied away from the organism (*in vitro*) reflect those of lysosomes in the living organism (*in vivo*). As an intermediate test system, lysosomes in living cells cultivated in controlled conditions in closed plastic dishes ('cell culture') can be used. Macrophages, derivatives of one of the white blood cell types, which as their name implies, are highly phagocytic (section 4.1), are often used. When macrophages take up materials which cannot penetrate membranes directly, endocytosis is involved (Chapter 4—see also Fig. 2–1). Having entered lysosomes the material is attacked by lysosomal enzymes. If it is digestible, for instance a protein, small molecular weight products (e.g. amino acids) which *can* penetrate the membrane are formed, and leave the lysosome, contributing to cellular nutrition (for instance, supplying amino acids for protein synthesis). Luckily, rates of production *in vivo* of these small molecules do not exceed the permeability potential of lysosomes: otherwise lysosomal breakage would result. However, if indigestible and non-penetrating, the compound accumulates in lysosomes, and causes the development of large numbers of enlarged lysosomes (Fig. 2–1). This development can be assessed by light or electron microscopy, and thus characterizes compounds which are non-digestible and non-penetrating.

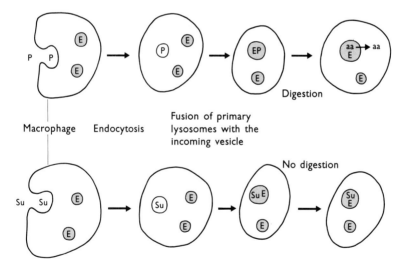

Fig. 2–1 Permeability of macrophage lysosomes. A comparison of the fates of endocytosed proteins (P) (which are digested to give amino acids (aa) which can diffuse out of the lysosome) and endocytosed sucrose (Su) (which is indigestible, and accumulates within lysosomes because it cannot penetrate the lysosomal membrane).

Foremost among such materials is sucrose, for which most animal cells possess no degrading enzyme. The lack of permeation of lysosomal membranes by sucrose in the living macrophage, validates the *in vitro* protection experiments described earlier. Observations on other compounds are also in close agreement in the two systems.

It is likely that lysosomal permeability does vary *in vivo*: for instance in some pathological states there seems to be rather increased permeability to small molecular weight substrates in living cells. This can be demonstrated by careful enzyme histochemistry. Many agents such as Vitamin A, some steroids, and some drugs are 'membrane-active': this indicates that they have a high affinity for membranes, to which they bind, and that such binding produces modifications in the properties of the membranes. Such agents may break membranes, or stabilize them, and often low concentrations of a particular compound will stabilize, while higher concentrations are destabilizing. Such stabilizing agents are important in attempts to treat malfunction of the lysosomal system, such as excess secretion of lysosomal enzymes (see sections 4, 6.4 and 7). In some circumstances, the stabilizers may inhibit the excess secretion, and in general may inhibit the membrane fusions involved in many lysosomal activities (section 3.4). An extreme case of membrane-active agents is represented by some particles with very high hydrogen bonding capacity (silica and asbestos, both industrial hazards, and urate crystals, which are present in the disease gout). After phagocytosis by cells, such agents disrupt the lysosomes from inside, by competing for and dissociating intra-membrane hydrogen bonding sites. Cell death usually ensues, as indicated already, emphasizing the importance of the lysosomal membrane for protecting the cell against autodestruction.

The lipid composition of membranes affects their stability, permeability and response to some membrane-active agents (which may bind specifically with a certain component). As mentioned already, lysosomal membranes, unlike other intracellular membranes, contain cholesterol, which favours stability of membranes. Thus another means of attempting to control pathological lysosomal hyperactivity is to change the lipid composition of the membrane. As most lipids are themselves membrane-active, they can mostly exchange between soluble and membrane-bound pools, thus traversing biological membranes by their own route involving actual incorporation (even if temporary) into membranes, rather than mere movement through membranes (as with sugars, amino acids, etc., discussed earlier). This is the route taken by the lipid products of complex lipid digestion in lysosomes, and as cholesterol shows very high exchange rates, it could probably be increased in lysosomal membranes relatively easily.

2.2 The intralysosomal pH

In view of the acid pH optima of most lysosomal enzymes, it might well be anticipated that the lysosome would maintain an acidic internal environment. Experiments with killed yeast cells carrying various indicators, adsorbed on their surfaces, suggested that within the lysosomes of living polymorphonuclear leucocytes the yeasts were indeed exposed to low pH. Since the yeast cells can be digested, the yeast surface is clearly an environment in which lysosomal enzymes are active.

In vitro measurements of pH within isolated lysosomes have been made recently. Because weak bases diffuse readily across biological membranes, mainly in the uncharged form (the charged forms having very low permeability) their equilibrium distributions between medium and lysosomal interior can be used to assess the pH of the latter. This is because on entering the lysosome, the weak base becomes protonated by a free H+ ion, and trapped within the lysosome because the protonated molecule has a positive charge and thus cannot permeate the membrane. Thus while the uncharged species will reach the same concentration on each side of the membrane, the charged form will be accumulated in lysosomes, and the relative (inside : outside) concentrations are determined by the relative pH values (providing all the internal molecules are in solution and not bound in any way). This is shown schematically in Fig. 2–2. Methylamine is concentrated from a very dilute solution up to 40–fold by lysosomes in 0.25 M sucrose at neutral pH, suggesting that the

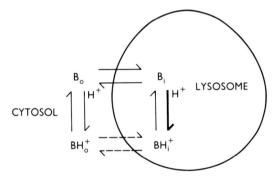

Fig. 2–2 The intralysosomal pH and the accumulation of weak bases. The permeating weak base (B) is in equilibrium with its protonated form (BH+), which permeates very poorly (as indicated by the dotted lines). If the H+ concentration within the lysosomes is higher than that in the cytosol, the BH+ form will reach much higher concentrations within than without. And thus the intralysosomal total base concentration ($[B_i]+[BH_i]$) will be greater than the cytosol ($[B_o]+[BH_o^+]$), as $B_o = B_i$. The ratio $\dfrac{[B_i]+[BH_i]}{[B_o]+[BH_o]}$ will increase as the intralysosomal [H+] rises, and can be used to determine the latter.

internal pH of the lysosome is approximately 1.6 units lower than that of the medium. This concentration is observed at 4°C or 37°C and is not dissipated by agents which induce H^+ (proton) permeability in biological membranes: it thus seems to be an equilibrium distribution, and not actively maintained. External monovalent cations can decrease the concentration by exchanging with intralysosomal protons. The consistency between observations under various conditions of the distribution of methylamine and cations (in low amounts) suggests that non-ionic binding does not complicate the distribution of methylamine, which is most easily explained by a proton gradient dependent on a Donnan potential across the lysosomal membrane.

The production of the proposed Donnan potential would require an intralysosomal concentration of fixed anions. Many such fixed anions have already been discussed (e.g. sialic acid on the internal surface of lysosomal membranes; acidic lipoproteins; excesses of acidic amino acids and phospholipids over their basic counterparts in the membrane); they may well be sufficient for the maintenance of the observed pH differential.

In contrast to the distribution of methylamine, several fluorescent dyes, such as the cationic amino-acridines, are passively accumulated by lysosomes by internal binding as well as protonation. This is shown by the inconsistency between distribution ratios of the different acridines, and by the maintenance of very high ratios even in the presence of high concentrations of external K^+, sufficient to abolish the concentration of methylamine. Concentration ratios of up to 500 are observed. One acridine orange-binding lysosomal component has been characterized.

Similarly, chloroquine, and other basic drugs and dyes are accumulated very considerably. Chloroquine reaches intralysosomal concentrations of 20 mM or more, and concentration ratios of several hundredfold. It has been suggested that this immense concentration indicates the occurrence of a lysosomal proton pump, pumping protons into the lysosome and providing internal protons to trap these large amounts of chloroquine. However, the observations on acridines described above show that protonation and binding, without the involvement of a proton pump, is probably sufficient.

There may be experimental underestimations of lysosomal acidity, and possibly *in vivo* contradiction of lysosomes might allow maintenance of greater acidity. In addition hydrolysis of ester or peptide bonds within lysosomes generates protons, which may contribute to the lysosomal pH. At present, therefore, there is no compelling evidence for a lysosomal proton pump: the intralysosomal acidity can be explained on the basis of Donnan equilibria alone.

3 Formation and Fate of Lysosomes

3.1 Synthesis of lysosomal components

There has been little work on the synthesis of lysosomal components. It is clear that synthesis of lysosomal proteins takes place on ribosomes in the normal manner, and that no protein synthesis occurs in lysosomes. Ribosomes on rough endoplasmic reticulum are involved. Indeed a rough endoplasmic reticulum fraction enriched in lysosomal enzymes has been obtained from rat kidney. It seems that there is coupling of formation of lysosomal vesicles and synthesis of lysosomal proteins, so that vesicles cannot be formed in the absence of newly-synthesized lysosomal proteins.

Rates of synthesis of β-glucuronidase increase during its hormonal induction in mouse kidney, while those of some other lysosomal enzymes seem unchanged. Rates of synthesis of several lysosomal enzymes are elevated in regressing mammary tumours, where lysosomal hyperactivity is involved. It is likely that changes in rates of degradation are involved in some of the situations where lysosomal enzyme activities change: control of both degradation and synthesis is known in the case of better-studied non-lysosomal enzymes.

There are several instances of such induction of increased lysosomal enzyme activities demonstrable *in vitro*. For instance, when presented with digestible phagocytosable substrates, cultured macrophages produce increased intracellular levels of most lysosomal enzymes. Endocytosis is necessary but not sufficient for this increase; the subsequent stages of the mechanism remain obscure. In some cases selective increases of groups of lysosomal enzymes, but not others, may occur (e.g. glycosidases, but not proteinases). And as already mentioned, even more specific changes may occur.

Elevated lysosomal enzyme levels are associated with many cancer cells, where they may be concerned with invasion of tissues, and also in cells of old organisms (or cells aged in culture) where perhaps they serve to remove defective proteins. Such proteins are made continuously in all cells by virtue of errors in the synthetic machinery, and are removed by degradation. As a defective protein involved in protein synthesis may catalyse the formation of larger numbers of defective proteins, the amounts of such proteins may accumulate in aged cells. Possibly the lysosomal increase in aged cells represents an attempt to deal with such increasingly defective proteins.

Even less is known about the incorporation of lipid precursors into lysosomal membranes, though phospholipids of lysosomes are degraded continuously. Similarly protein incorporation into lysosomal membranes has not been studied.

3.2 Translocation of lysosomal components and production of primary lysosomes

After formation of lysosomal proteins on ribosomes attached to the endoplasmic reticulum, they are transported probably *via* the cisternal space of the endoplasmic reticulum to smooth microsomes, then to the Golgi apparatus, and finally to lysosomes. Contributions from other routes cannot be excluded at present. It is not clear whether the Golgi cisternae or only the peripheral vesicles are involved. Subfractionation of Golgi preparations shows concentration of lysosomal enzymes in vesicles rather than dictyosomes (Fig. 3–1), but it is clear that the first vesicles containing lysosomal enzymes (the 'primary lysosomes') are formed in the vicinity of the Golgi apparatus.

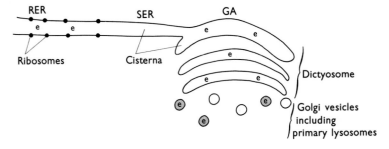

Fig. 3–1 Synthesis and transport of lysosomal enzymes. Lysosomal enzymes (e) are synthesized on the rough endoplasmic reticulum (RER) and enter the cisterna. They travel via the smooth endoplasmic reticulum (SER) to the cisternae of the Golgi Apparatus (GA) dictyosome. The dictyosome cisternae are probably in direct continuity with each other, in a three-dimensional network, and the lysosomal enzymes can thus travel from one face of the stack to the other. At the mature face, varied Golgi vesicles are formed and these include primary lysosomes.

Lysosomal enzymes are mostly glycoproteins, and the carbohydrate moieties of such proteins are added sequentially after protein synthesis, in various parts of the endoplasmic reticulum and Golgi apparatus. As some of the enzymes involved (glycosyl transferases) are almost entirely restricted to the Golgi apparatus, this also implies that lysosomal enzymes pass through the Golgi apparatus after formation. As already mentioned, changes in charge and solubility accompany the progressive glycosylation of lysosomal enzymes.

This translocation route for lysosomal enzymes confirms earlier histochemical work showing that phagocytic vesicles often first become acid phosphatase-positive in the Golgi area, presumably by fusion with primary lysosomes, and that several acid hydrolases can be demonstrated in the Golgi apparatus. In addition it supports the pathway of transport of lysosomal proteins in macrophages suggested on the basis of autoradiography of total proteins synthesized by macrophages in the presence of radioactive amino acids, during induction of lysosomal enzymes.

An alternative scheme of lysosomal enzyme transport is embodied in the GERL (Golgi associated endoplasmic reticulum forming lysosomes) hypothesis of Novikoff. This proposes direct formation of lysosomes from smooth endoplasmic reticulum close to the Golgi apparatus but without Golgi involvement. Budding of endoplasmic regions rich in lysosomal enzymes is envisaged. In addition, pairs of endoplasmic reticulum membranes participating in autophagy (see section 5) may form vesicles with two membranes containing lysosomal enzymes between the membranes, the inner of which is then digested. There is histochemical and morphological evidence for such routes (Fig. 3–2).

The heterogeneity of enzymatic composition of lysosomes (see sections 1.4 and 3.3) even in single cell types implies that lysosomal packaging is

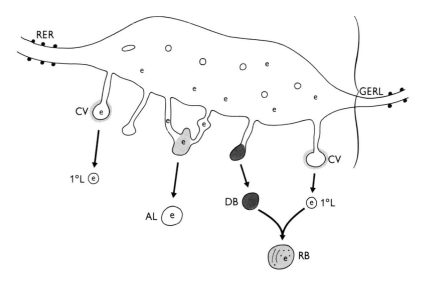

Fig. 3–2 The GERL system. The GERL is a specialized region of smooth endoplasmic reticulum, and probably gives rise to three types of lysosome. Firstly, the coated vesicles (CV) which probably form primary lysosomes (1°L). Secondly, via a type of autophagy, it produces autolysosome (AL). Thirdly, dense bodies (DB), which may contain lysosomal enzymes (e) initially, and which can also fuse with other lysosomes, eventually giving rise to residual bodies (RB), containing some remnants of the digestive processes. RER: rough endoplasmic reticulum.

not a uniform process, but probably depends on the relative amounts of individual enzymes available at a given time, and on more subtle factors. Precedents for remarkable functional differentiation in the Golgi apparatus are known. For instance, in the white blood cell the polymorphonuclear leucoctye, the lysosomal ('azurophil') granules, and a second granule ('specific') are formed from opposite faces of the Golgi apparatus, and in spermatozoa, two different types of lysosome, the acrosome and the cytoplasmic droplet, are formed at different times from the Golgi apparatus.

3.3 Transformation of lysosomes

The newly formed primary lysosome, which contains lysosomal enzymes but little substrate, may undergo a wide array of fusion processes, so joining the functional secondary lysosome population, which contains plenty of substrate. Fig. 3–3 is an overall scheme of lysosomal fusions and lists standard (and alternative) terms for the various components of the system.

In order to fuse, intracellular vesicles have to move within the cell, become juxtaposed, and undergo the complex process of fusion itself. We next consider these processes, before describing the various fusions actually occurring in the lysosomal system.

The saltatory movements of intracellular vesicles in macrophages depend on the presence of intact microtubules, the linear tubular aggregates in cytoplasm like those of flagella and cilia. Some evidence

Fig. 3–3 The general categories of primary lysosomes (containing lysosomal enzymes, but not having received substrates), and secondary lysosomes (in which substrates are undergoing enzymatic attack) are distinguished. In addition, the term prelysosome covers PiV, PhV and AP: vesicles containing substrates but awaiting the entry of lysosomal enzymes. Several alternatives for the terms used here are extant (see DE DUVE and WATTIAUX, 1966), the most important are: autophagic vacuole or cytosegresome, which cover AP and AL together; and endocytic vesicle and heterophagosome which cover PiV and PhV together.

It should be emphasized that not all the lysosomal transformations indicated here necessarily occur in every cell type, and that some lysosomal activities are omitted (e.g. penetration of the nuclear membrane). But in general there is a remarkably continuous exchange between all members of the lysosomal system: thus residual bodies can still fuse with most other types of lysosome, and there is free fusion between vesicles produced by autophagy and those from heterophagy.

Abbreviations: Within membrane: e, lysosomal enzymes; p, secretory proteins; s, soluble substrates for lysosomal digestion; particulate substrates are represented as solid areas. *Outside membranes:* RE, rough endoplasmic reticulum; SE, smooth endoplasmic reticulum; GA, Golgi apparatus; SV, secretory vesicle; AP, autophagosome; AL, autolysosome; PE, peroxisome; MVB, multivesicular body; RB, residual body; PiV, pinocytic vesicle; PhV, phagocytic vacuole; HL, heterolysosome; MI, mitochondrion.

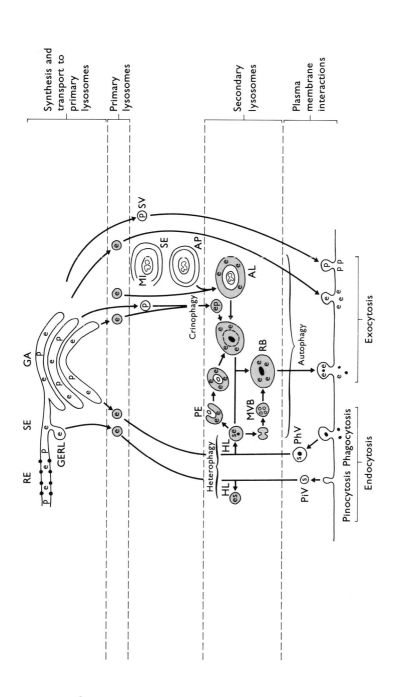

suggests that the intracellular movements of lysosomes may also be restrained by the peripheral band of a second linear structure, the contractile microfilaments present in most cells, particularly in respect of access to the plasma membrane, but a positive role of microfilaments is also possible, at least in moving lysosomes towards the periphery of the cell during exocytosis (section 4.3).

To what degree the frequency of intracellular vesicle fusion is controlled by their rates of movement is unclear. Colchicine (a microtubule disaggregator) does not retard digestion of previously phagocytosed substrates by macrophages, and thus seems not to reduce fusion of primary lysosomes with phagosomes. This indicates that even when microtubules are disrupted the intracellular movement of lysosomes is sufficient for the required frequency of fusion. It seems more likely that factors intrinsic to the vesicle membranes control frequency of fusion. This can be studied by means of model membranes (liposomes). Liposomes are concentric lipid bilayers (see Study in Biology no. 27) enclosing aqueous spaces (Fig. 3-4) which form spontaneously when dry lipids are allowed to swell in acqueous solutions. Multilayer structures form first, but these can be converted to unilamellar structures by sonication. It is found that composition of the liposome membranes affects their frequency of fusion with each other *in vitro*, and with plasma

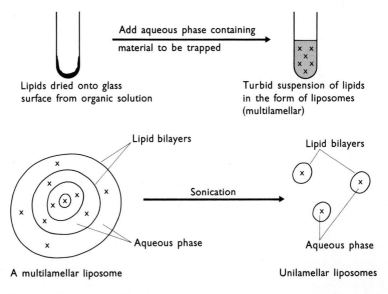

Fig. 3-4 Formation of liposomes, containing a non-penetrating solute (x). On addition of the aqueous phase, the liposomes form spontaneously, and trap some aqueous fluid containing x. Molecules of x which remain outside the liposomes, in the bulk phase, may be removed by dialysis or gel filtration.

membranes of intact cells in culture. This is consistent with the effects of lipid composition on membrane stability, as discussed earlier.

Frequencies of fusion of lysosomes with other vesicles cannot easily be measured *in vivo*, though the isolation of phagosomes containing the low density material, oil red O, from macrophages and neutrophils allows estimation of the amount of lysosomal enzyme released into the phagosome by the time of preparation. Using this method it was shown that the membrane stabilizer cortisone, the microtubule disaggregator colchicine, and cyclic nucleotides (cellular modulators: see section 4) have little effect on fusion of lysosomes and phagosomes.

Lysosomes fail to fuse with certain endocytotic vesicles: for instance those containing live *Mycobacterium tuberculosum* and certain other microorganisms. The drug suramin, which is used to kill invading trypanosomes, is accumulated in lysosomes and apparently inhibits fusion of lysosomes in macrophages with phagosomes containing killed yeast, while the lectin concanavalin A similarly inhibits macrophage phagosome-lysosome interaction. The continuing presence of the agents is required for inhibition, which is reversed upon removal. Lysosomal membranes do not fuse with the membranes of other intracellular organelles (nuclei, mitochondria, peroxisomes) except those of the endoplasmic reticulum participating in autophagy (see section 5). In contrast to this fairly discriminating pattern of normal fusions the perplexing Chediak-Higashi syndrome seems to involve exaggerated non-selective fusions, resulting in enlarged lysosomes in some cell types, though in others fusion of primary lysosomes with phagosomes may be delayed. In addition there is some evidence from serial sectioning of rat liver tissue and from a study of hamster liver lysosomes for permanent continuities between some types of lysosomes and elements of endoplasmic reticulum. The mechanisms involved in membrane fusion are little understood, but, as fusion of lysosomes has been observed *in vitro*, approaches to the study of factors controlling such fusions and their specificity are open.

As Fig. 3-3 indicates, primary lysosomes normally fuse with endocytotic vesicles of all kinds, and with autophagic vacuoles (containing diverse intracellular material). Phagocytosed tracer particles can permeate the whole lysosomal system, proving that most lysosomes can fuse with each other.

In certain systems (preputial gland, uterus) vacuoles containing target-specific hormones can be formed, presumably by pinocytosis, fuse with primary lysosomes, and be transported by them to the nucleus. Electron microscopy has revealed invagination of the nuclear membrane to allow entry of these vacuoles and thus it is tempting to speculate that they may be an important vector for these hormones, and that this lysosomal function may be required for hormone action. Perinuclear clustering of lysosomes had been described previously in several systems, and it has been suggested that lysosomal enzymes acting on nuclear materials might be involved in conversion ('transformation') of cells to the cancerous state, although there is still little evidence of this.

This diverse pattern of lysosomal fusions is partly responsible for the heterogeneity of lysosomal composition mentioned already. The end product of lysosomal fusions, the 'residual body', contains indigestible material, such as lipofuscin, which is a complex lipid compound, and it was suggested that some residual bodies were inert and lacked active lysosomal enzymes (the 'post lysosome'). However, it is found that residual bodies containing particles of thorium dioxide, given four weeks previously, do fuse with newly formed phagocytic vesicles containing gold particles, and do contain active acid phosphatase. Thus the postlysosome remains an unsubstantiated theoretical concept.

3.4 Cytoplasmic liberation of lysosomal contents

As indicated already lysosomal enzymes are not normally found in the cytosol. However, they may be liberated into the cytosol in some pathological conditions. Usually this is a late change in the degeneration of cells, and initiation of cell death has occurred long before lysosomal breakage; but in some special cases, such as photodynamic damage to sensitized cells, and in ultraviolet irradiation, lysosomal leakage may well initiate cell death. Other examples are cell damage caused by silica, asbestos, urate crystals, beryllium, viruses and possibly some carcinogens. Leakage is often demonstrated, but it should be noted that some of these results are called into question by experiments in which lysosomes are preloaded (by endocytosis) *in vivo* with electron dense particles. After the cell damage has been induced, lysosomes are studied in the electron microscope. This approach concentrates on liberation of enzyme (reflected by the particles) rather than permeability changes (reflected perhaps by histochemical activation) since the latter may not be pathologically important. In several cases no release of particles to the cytoplasm is observed though histochemical activation has been claimed previously. It is notable that some tissue cells contain cytosol proteinase inhibitors, which would limit damage caused by lysosomal enzymes released into the cytosol.

3.5 Degradation of lysosomal constituents

It has been shown that proteins of liver lysosomal membranes and soluble phase are heterogeneous in their rates of degradation. This observation excludes simultaneous degradation of whole lysosomes or exocytosis of liver lysosomes as major fates of the lysosomal proteins. The time for degradation of half the molecules of a protein originally present (during which time the original molecules will have been replaced by new ones) is referred to as the half-life of the protein, and is the usual measure of protein degradation. The average value for the proteins of whole liver lysosomes is 30 h, while the soluble lysosomal proteins have a half-life of 24 h and the membrane proteins 29 h, indicating rapid turnover. There is the question of whether lysosomal proteinases degrade other lysosomal enzymes (including proteinases). Of course, observed enzyme levels are the result of both supply (from synthesis and vesicle fusion) and degradation, and these rates are largely unknown. The enzymes have presumably evolved to be stable at acid pH and resistant to the lysosomal proteinases: their excellent survival during autolysis *in vitro* (when a tissue is homogenized and incubated) and ischaemia *in vivo* (when a tissue is depleted of oxygen) is consistent with this.

Lysosomal enzymes are probably also protected from digestion by virtue of their intra-lysosomal distribution (membrane bound?), while foreign materials (including lysosomal enzymes isolated from other species) do not have this advantage. Therefore probably some slow intralysosomal degradation of endogenous enzymes occurs, and this is consistent with evidence that lysosomes are a major site of degradation of intracellular proteins in general (section 6.2) and that lysosomal components do not share a common half-life.

The rapid degradation of sugar moieties in total lysosomal glycoproteins has been mentioned; but there seems to be little information on degradation of phospholipids and other components of lysosomes.

4 Endocytosis and Exocytosis

4.1 Phagocytosis

Endocytosis is a general term for processes by which external material enters a cell in vesicles formed from the plasma membrane. When the endocytic vesicle contains particles (such as bacteria, thorium dioxide, etc.) it is termed a phagosome or phagocytic vacuole (Fig. 4–1 and Fig. 4–2). Phagocytosis can be dissociated into an adsorption and an entry phase. The process is sometimes facilitated by coating of the particle with serum components (such as immunoglobulins) for which the phagocytic

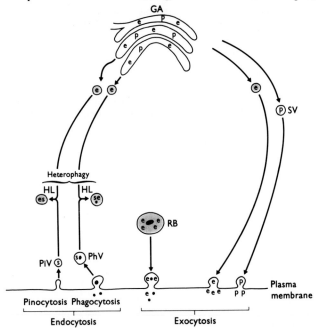

Fig. 4–1 Endocytosis and exocytosis. Exocytosis of secondary (RB) and primary lysosomes is shown, together with that of secretory proteins (p) in specialized secretory vesicles (SV). Examples of such proteins are serum proteins, produced by the liver, and some protein hormones produced by endocrine glands. *Abbreviations: Within membranes:* e, lysosomal enzymes; p, secretory proteins; s, soluble substrates for lysosomal digestion; particulate substrates are represented as solid areas. *Outside membranes:* GA, Golgi apparatus; HL, heterolysosome; PhV, phagocytic vacuole; PiV, pinocytic vesicle; RB, residual body; SV, secretory vesicle.

cell (macrophage, polymorphonuclear leucocyte) has receptors, surface molecules capable of binding the serum components specifically. Not all particles require such 'opsonization' for rapid uptake (e.g. multilamellar liposomes). Obviously, for a unicellular organism which relies upon phagocytosis for its main nutrient source, very wide uptake capacities are required, and thus an opsonizing process is not usually involved.

Mechanisms of entry vary with particle size, and seem to require energy and the active participation of micro-filaments, the contractile intracellular filamentous structures. Rapid fluxes of internal calcium stores seem to be involved, and calcium may effect the microfilament contraction. The lectin concanavalin A inhibits both adhesion and entry of some particles into mouse macrophages and also shows some specificity in inhibition of phagocytosis by polymorphonuclear leucocytes. This may be due to selective binding to particular membrane receptors involved only in the uptake of certain particles.

The membrane involved in phagocytosis does not usually contain all membrane constituents: for instance, little internalization of various amino acid transport sites occurs. On the other hand, some membrane receptors, such as those for concanavalin A, may be selectively included in the phagosome. Both these types of selectivity (inclusion and exclusion) seem to depend on microtubule function, as they are blocked by colchicine, the microtubule disaggregator. Thus a transmembrane function of microtubules has to be considered; indeed it is known that microtubules can interact physically with membranes.

4.2 Pinocytosis

Two apparently distinct processes are subsumed under this term. Both involve uptake of external solution, but in macropinocytosis, the vesicles formed are over 300 nm in diameter, and microfilaments are involved, whereas in micropinocytosis, vesicles are smaller (only visible by electron microscopy) and microfilaments probably are not involved.

Components of the medium may be taken up in fluid or adsorbed to the membrane of the forming vesicle, and the fractional uptake of adsorbed materials can vary considerably and be much greater than that of soluble material (Fig. 4–3). Several anionic compounds (large and small molecules) stimulate pinocytosis by macrophages. Stimulation by polycations has been observed in some systems, while cortisone and other stabilizers such as chloroquine often inhibit pinocytosis. The dye Trypan Blue has a remarkable biphasic effect on pinocytosis by yolk sac, and the wide range of rates reproducibly obtainable with this agent at least suggests the possibility that there might be subtle cellular control over the process.

It appears that some lysosomal enzymes can be taken up by fibroblasts (connective tissue cells often obtained by outgrowth from pieces of skin)

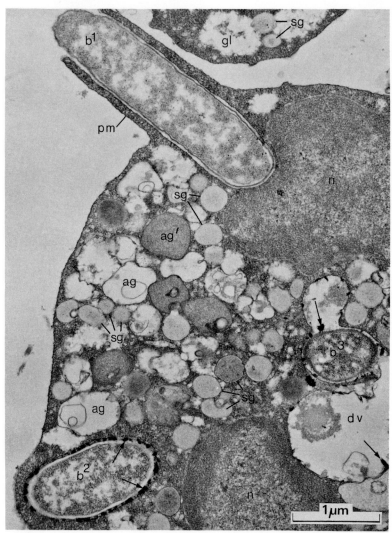

Fig. 4-2 Rabbit exudate polymorphonuclear leucocyte (PMN) (30 s after exposure to live *Escherichia coli*) reacted for alkaline phosphatase. The cell has phagocytosed three bacteria (b^1–b^3), which are visible here at different stages of ingestion. One microorganism (b^1) is in the process of being engulfed, with the plasma membrane (pm) of the PMN forming the wall of the phagosome. Note the lack of reaction product around b^1. A second bacterium (b^2) has been internalized, and alkaline phosphatase reaction product is apparent in the narrow space between the vacuole membrane and the bacterium (arrows), indicating that

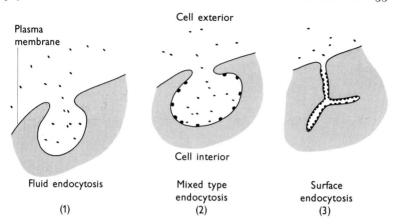

Fig. 4-3 Selectivity in pinocytosis. Molecules may enter either in solution (1) or bound to the surface (3), or by a combination (2). Process (3) can allow entry of a much larger proportion of external molecules per unit time than process (1), and the proportion in (3) varies from molecule to molecule, according to the degree to which the molecule binds to the plasma membrane. In contrast, all molecules entering a cell by process (1) enter at the same proportional rate, i.e. non-selectively. Thus, the combination of processes 1–3 acting on a mixture of molecules allows considerable diversity in proportional entry rates.

in culture by adsorptive pinocytosis (i.e. bound to the plasma membrane), and that this depends on the presence of carbohydrate moieties on the enzymes. The process may be similar to that resulting in rapid uptake and catabolism of desialylated glycoproteins by liver parenchymal cells, involving a plasma membrane receptor (section 6.1). Selective uptake of some lysosomal enzymes by fibroblasts was prevented by chloroquine (5×10^{-6} M) although pinocytosis of dextran was unaffected. The physiological significance of this selective process is discussed in the next section, and its possible therapeutic application subsequently. As with phagocytic vacuoles, not all pinocytotic vesicles fuse with lysosomes; particularly in epithelial cells they may be exocytosed (see section 4.3) after transporting their contents across a cell

---

the contents of specific granules (sg) have been discharged into this phagocytic vacuole. A third bacterium (b^3) is present (in the lower right portion of the micrograph) within a larger digestive vacuole (dv). Some enzyme reaction product can also be seen in this vacuole (arrows). The adjacent specific granules contain no enzyme deposits, nor do the azurophils, which appear to be completely (ag) or partially (ag') extracted by the procedures for enzyme demonstration. Other empty-appearing areas, best seen in the cell above, probably represent islands of glycogen particles (gl) which were extracted by the preparatory procedures. By kind permission of Dr D. F. Bainton, and North Holland Publishing Co.

without joining the lysosomal system (the process of diacytosis). The transfer of immunoglobulins from mother to foetus in rats seems to be an example of this process.

4.3 Exocytosis

A reversal of endocytosis results in release of the contents of an intracellular vesicle to the extracellular space (Fig. 4–1) without loss of cytoplasmic components (for which lactate dehydrogenase is a useful marker). The ability of normal fibroblasts to correct the lysosomal enzyme deficiency of mutant fibroblasts (characterized by loss of a single lysosomal hydrolase) during co-cultivation suggests that lysosomal enzymes may normally be secreted by fibroblasts, and then endocytosed by the same or other cells. This phenomenon has not yet been demonstrated directly. But when fibroblasts are treated with cytochalasin B which prevents microfilament function, pinocytosis is inhibited and the cells undergo a nett loss of lysosomal enzymes to the medium. Chloroquine produces the same loss, without inhibiting general pinocytosis.

These results are relevant to the human 'I-cell disease' in which there is a deficiency of many lysosomal hydrolases. Cultured 'I-cells' (fibroblasts) have low activities of many hydrolases, while the media have correspondingly high activities. It is unclear whether the lesion in I-cells causes excess secretion, or poor uptake of I-cell enzyme. Either might be a consequence of formation of abnormal enzymes, perhaps due to a common defect in their synthesis (glycosylation?) or to a general abnormality of cellular membranes. Thus it is not decided yet whether secretion of lysosomal enzymes followed by re-uptake is a normal physiological mechanism for integration of fibroblast lysosomal enzymes into functional lysosomes. In principle, the process seems grossly inefficient, yet may be a real alternative to the mechanisms outlined above, at least in those cells showing the intracellular accumulations of undegraded material which characterize the disease (many cell types are spared).

Enzymes from normal lysosomes are exocytosed by some tissues while residual bodies may be 'defaecated' from tissues including liver and from amoebae, the Protistan *Tetrahymena* and other cells. Release of endocytosed latex particles has not been seen in long-term studies of synovial fibroblasts of macrophages or in several other systems, though secretion of lysosomal enzymes from these cells does occur. Some cells do not defaecate their residual bodies, and in one such case, a ciliate of the genus *Tokophrya*, cell death is associated with the presence of numerous engorged vacuoles, and can be hastened by over-feeding.

Exocytosis of lysosomal enzymes from connective tissues in culture can be induced by such agents as the membrane active substance vitamin A, or the indigestible sugar sucrose which causes lysosomal distension and accumulation (in cartilaginous limb bone rudiments), and by parathyroid hormone acting on bone organ cultures. Many of the inducing agents are membrane destabilizers, and the processes can often be inhibited by stabilizers such as steroids.

At present, however, phagocytic cells are the main system for study of secretion of lysosomal enzymes. It was noted in 1963 that phagocytosis may be accompanied by selective release of lysosomal constituents, and early work on macrophage neutral proteinases also suggested this. Frequently release is due to fusion of primary lysosomes with forming phagosomes before they are completely closed. In addition, secretion can be dissociated from uptake in some systems, for instance when normally phagocytosable substrates are present on large solid supports ('frustrated phagocytosis') or when uptake is prevented by cytochalisin B (a microfilament inhibitor): see Fig. 4-4.

It has been suggested that particulate agents producing long-lasting ('chronic') inflammation may exert part of their effects by inducing lysosomal enzyme release from macrophages, and it has been shown that certain non-inflammatory particles do not cause release *in vitro* while inflammatory particles do. Several anti-inflammatory steroids decrease release, while inactive steroids do not. Thus it seems probable that released lysosomal enzymes are important agents of inflammation.

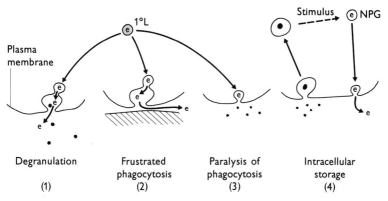

Fig. 4-4 Exocytosis of lysosomal and other enzymes accompanying phagocytic processes. In mechanisms 1, 3, 4, normal particulate stimuli are applied. In 1 degranulation of primary lysosomes into the forming phagosome occurs to some extent before it has closed. In 2 and 3, phagocytosis is initiated, but not completed: in 2 this is because a phagocytosable material has been insolubilized on a large solid support too big to enter the cell, and in 3 because phagocytosis has been inhibited, for instance by cytocholasin B. In 4, intracellular storage of non-digestible material somehow stimulates the release of neutral proteinases from probably non-lysosomal granules. *Abbreviations:* 1°L, primary lysosome; e, enzyme; NPG, neutral proteinase containing granule.

Phagocytosis of latex by synovial fibroblasts leads to a large and sustained increase in the rate of release of some neutral proteinases accompanied by a smaller proportional increase in release of lysosomal acid hydrolases. Since the secretion continues while latex is stored by the cells, and while no further phagocytosis is occurring, it probably involves dissociation of site of release (of the neutral enzymes, which may be in granules distinct from lysosomes) from the site of phagocytosis. This dissociation has also been suggested for lysosomal enzyme release from macrophages. There are differences in timing of degranulation of polymorphonuclear leucocyte specific and azurophil (lysosomal) granules into phagosomes, and selective extracellular release of these separate populations can be achieved. Thus in several systems differential release of lysosomal and non-lysosomal enzymes can occur, and the importance of enzymes with neutral pH optima, often originating from the non-lysosomal vacuoles, is discussed later.

Immune complexes (formed by the reaction of antibodies with foreign materials which elicited their formation) and other serum components, especially when in particulate form, also often cause secretion. Induction of secretion by these agents in solution (rather than in particulate form) requires rather specialized conditions. They are generally much more effective when the molecules or aggregates have several binding sites for cell membrane receptors than when they are converted experimentally into forms with only one such site. This may be because in the former state they can link together several molecules of receptors into a cell surface cluster; such clustering of membrane components may facilitate membrane fusion processes, both in endocytosis and exocytosis. In the latter state, such cross-linking cannot occur.

Microtubules seem to be involved in exocytosis, as colchicine often inhibits. The opposed effects of cyclic-AMP (inhibitory) and cyclic-GMP (activatory), deduced from indirect experiments on externally supplied cyclic nucleotides and inhibitors of their hydrolysis, may be due to effects on protein kinases which phosphorylate components of microtubules and so modify their activity. Many examples of control of activity of enzymes by phosphorylation by protein kinases (using cyclic nucleotides) and dephosphorylation by phosphoprotein phosphatases, are now known. So such effects on microtubules may well be demonstrated shortly. Intracellular levels of cyclic nucleotides have been measured during the early stages of phagocytosis and release in polymorphonuclear leukocytes: cAMP falls and cGMP rises. There is an early calcium influx, and experimental production solely of a calcium influx is sufficient to induce secretion in several systems.

It seems that as in many systems the cyclic nucleotides may be 'second messengers' in transmitting the initial membrane stimulus, leading towards the final membrane fusion, and exocytosis. The idea that the two major cyclic nucleotides, cAMP and cGMP have opposed effects on all cellular processes (which has been dubbed the 'Yin-Yang' hypothesis, referring to ancient philosophical concepts concerning antithesis and unity) is supported by results in this system. cAMP seems to be an inhibitor of secretion, while cGMP is an activator. Thus normal secretion is accompanied by increased intracellular cGMP and decreased cAMP; and experimental introduction of calcium mimics these effects on intracellular cyclic nucleotides. Direct evidence of modification of microtubules by cyclic nucleotides is still limited. Figure 4–5 presents a minimum scheme for intermediates in exocytosis: it presents no suggestion as to the mechanism of membrane fusion, about which we still remain abysmally ignorant.

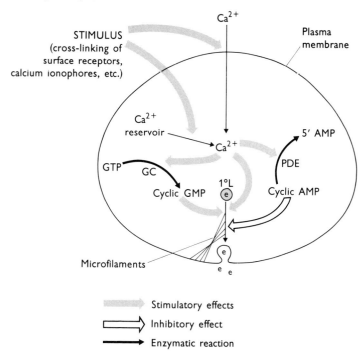

Fig. 4–5 A simplified scheme of the mechanism of induction of exocytosis. Calcium influx or release from an intracellular reservoir is consequent upon interaction of the stimulant with the plasma membrane. The resultant rise in intracellular calcium concentration initiates exocytosis and facilitates production of cyclic GMP by guanyl cyclase (GC), and removal of cyclic AMP by phosphodiesterases (PDE). Cyclic GMP stimulates the exocytosis, while the inhibitory effect of cyclic AMP is minimized by its breakdown. Mechanisms for recovery of calcium and cyclic nucleotide levels to normal are not indicated. *Other abbreviations:* 1°L, primary lysosome; e, lysosomal enzymes.

5 Autophagy and the Accumulation of Materials

5.1 Autophagy

Intracellular components may enter lysosomes by two known routes comprising autophagy: either the material (often intact organelles) is surrounded by pairs of smooth endoplasmic reticulum membranes which fuse to form double-membrane bounded vacuoles; or a lysosome may invaginate and enclose a piece of cytoplasm in a similar (but normally smaller) double-membrane bounded vacuole (see Figs. 1–6 and 1–7). In both cases the vacuole may contain lysosomal enzymes by virtue of its origin but with the first mechanism this may not always apply, and enzymes may be supplied by incoming primary lysosomes. The inner membrane of the pair normally disappears rapidly so that the cytoplasm or organelle within becomes available for digestion. A variant of the second mechanism is probably mainly responsible for the formation of multivesicular bodies—several small invaginations occur to form a vacuole containing several small vesicles. This variant may perhaps be important in degradation of lysosomal membrane components, since it results in the internalization of membrane within the lysosome, which may progressively remove the constraints on degradation discussed earlier.

The autophagic vacuoles thus apparently contain random samples of cytoplasmic material or intact organelles together with cytoplasmic material. However, selective uptake may occur (see section 6.2) and selective autophagy of endoplasmic reticulum and soluble materials or mitochondria and ribosomes is known in some circumstances. The various forms of autophagy are represented in Fig. 5–1.

Autophagy is most common in organs undergoing physiological (such as the uterus after delivery) or pathological regression, and programmed cell death (as in certain insect muscles during metamorphosis), but it is also a normal occurrence in non-regressing tissues. This is also implied by the intralysosomal accumulation of glycogen (a cytosol component) observed in certain storage diseases where cytoplasmic glycogenolytic systems are intact, but lysosomal systems for degrading glycogen are genetically defective. Glucagon treatment of liver leads to increased autophagy judged morphologically, and several other inducers of autophagy in certain tissues or cells are known (chloroquine, hyperoxia, starvation). Increased endocytosis is often accompanied by accentuated autophagy, which may serve to remove excess internalized membrane.

Only the most obvious instances of autophagy (such as mitochondrial

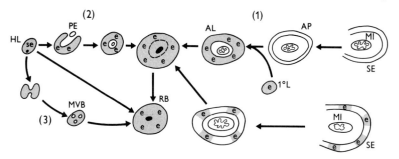

Fig. 5–1 Autophagy. Three main varieties are shown. In (1), endoplasmic reticulum (SE) regions surround organelles, and present them to the lysosomal system. The SE regions may or may not contain lysosomal enzymes. Microautophagy is a specialized form of (1). In (2) a heterolysosome (HL) surrounds and internalizes an organelle, while (3) results primarily in the internalization of lysosomal membrane, often forming multivesicular bodies (MVB). *Abbreviations: Within membranes:* e, lysosomal enzymes; s, soluble substrates for lysosomal digestion; particulate substrates are represented as solid areas. *Outside membranes:* SE, smooth endoplasmic reticulum; AP, autophagosome; AL, autolysosome; PE, peroxisome; MVB, multivesicular body; RB, residual body; HL, heterolysosome; Ml mitochondrion; 1°L, primary lysosome.

uptake and digestion) are easily identifiable by electron microscopy. More subtle variants, such as microautophagy, which is essentially very small-scale autophagy, are difficult to detect. In addition, although the volume of the autophagic vacuole system of a tissue at a particular time may be roughly estimated, there is presently no morphological means of determining the lifetime of such vacuoles. As a result, there are no meaningful estimates of the rate of autophagy in any system.

Figure 5–2 illustrates a variant of autophagy (crinophagy) which occurs in several types of secretory cell, such as hormone producing cells. In this process excess secretory vesicles are removed by fusion with lysosomes and subsequent degradation. It is distinct from autophagy in that the vesicles to be removed fuses with lysosomal membranes, rather than being surrounded by membranes. This may be an important mechanism controlling rates of secretion and can be induced in some systems by blocking secretion. It may also be far more widespread than presently realized: for instance it may occur continuously in liver, modulating the rate of secretion of serum proteins.

5.2 Accumulation of materials by lysosomes

Four main mechanisms of uptake of materials into lysosomes have already been discussed: phagocytosis (particulate matter), pinocytosis

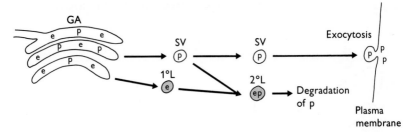

Fig. 5–2 Crinophagy. A secretory vesicle (SV) containing secretory proteins (p), may achieve exocytosis, or, particularly in conditions of high rates of production, undergo fusion with primary lysosomes (1°L), and subsequent degradation. This process may be also accentuated when exocytosis is blocked. *Abbreviations: Within membranes:* e, lysosomal enzymes; p, secretory proteins. *Outside membranes:* GA, Golgi apparatus; 1°L, primary lysosome; 2°L, secondary lysosome; SV, secretory vessel.

soluble and adsorbed matter), autophagy (intracellular materials) and permeation. Each of the processes can result in uptake of both digestible and indigestible compounds and the latter may thus accumulate. Many low molecular weight compounds, including drugs, dyes and carcinogens, are accumulated after rapid permeation, probably by protonation of the bases, and hydrophobic interactions. On the other hand, cationic macromolecules, and probably particulate and complexed extracellular metals, can enter by slower endocytosis, and cytoplasmic components (such as protein-bound metals) may well be scavenged by selective autography (section 6.2). Thus a very wide array of materials can be accumulated in lysosomes if they become available; some examples and the probable mechanisms of their accumulation are given in Fig. 5–3.

Fig. 5–3 Accumulation of materials by lysosomes

Mechanism of accumulation	Examples of materials so accumulated
Phagocytosis	Iron (particularly in the disease, haemochromatosis); Silica (in the disease, silicosis); Asbestos (in the disease, asbestosis). Indigestible fragments of microorganisms.
Pinocytosis	Sucrose, Mannitol, some dyes. Indigestible macromolecules such as polysaccharides, synthetic polymers (e.g. Triton WR-1339). In lysosomal storage diseases, extracellular mucopolysaccharides, etc.
Autophagy	Intracellular materials, particularly in lysosomal storage diseases.
Permeation	Protons; weak bases; some dyes, drugs and carcinogens.

6 Functions of Lysosomes

The previous chapters have described the discovery of lysosomes and the subsequent elucidation of their properties. That is their structural organization, distribution, physico-chemical properties, life cycle within the cell, and many of their mechanisms of action. In the course of this many of their actual functions have been touched upon. Now, in the final chapter, their functions will be described both specifically, and in the context of the biology of the whole cell or organism. The extreme diversity of these functions (see particularly section 6.4) raises many interesting questions which require much further research.

6.1 Degradation of endocytosed materials

This is the classic function of phagocytic cells, which are usually rich in lysosomes. Removal of foreign particles (for instance bacteria and viruses) by such phagocytes in multicellular organisms is an important protection against infection; in unicellular organisms phagocytosis is usually a major nutrient source, rather than simply a protective event. The mechanism of killing of organisms by phagocytes is still unclear, but permeability changes in the organism, leading to loss of metabolites and cell death, often occur before significant proteolysis. In polymorphonuclear leucocytes, the oxygen-dependent systems for generation of H_2O_2 and superoxide radicals seem to be of vital importance in killing since genetic deficiency of the systems greatly decreases resistance to infection, in spite of the presence of the bactericidal cationic proteins. In addition, dead cells originating from tissues of the animal itself can be endocytosed by phagocytes and degraded in lysosomes. This is part of normal cellular turnover, as many cells have quite brief lifespans. As mentioned already, lysosomes do not seem to be involved in initiating either normal, or most forms of pathological, cell death. In situations of physiological remodelling or pathological regression, besides (or sometimes instead of) an induction of autophagy in the regressing cells, there is often an invasion of phagocytic cells such as macrophages, which remove dead cells, and subject them to lysosomal digestion. Figure 6-1 lists several situations of physiological cell death and indicates the degree of involvement of internal lysosomal activity and invading phagocytes.

Many extracellular substrates are degraded after endocytosis. For instance, digestion of connective tissue components, which is a normal process, becoming pronounced during remodelling, involves two stages.

The first is an extracellular partial digestion (section 6.4) which produces soluble components from the solid connective tissue material. Such components, undergo modifications which accelerate their endocytosis allowing lysosomal digestion. Into this category fall serum carnivorous plants and in many fungi (whose heterotrophic nutrition involves the use of external organic material).

System	Intracellular lysosomal activity including autophagy	Activity of phagocytic cells in endocytosing large cell debris
	(Relative importance)	
Embryonic		
Sympathetic nerve ganglia	0	+
Developmental and Hormonal		
Arthropod muscles and nerves	+ +	0
Tadpole tail	+ +	+ + +
Some Annelid tissues	0	+ + +
Pathological		
Lethal Anoxia	0	+ + +

Fig. 6–1 Modes of destruction of cells during physiological and pathological degeneration. Rough quantitation is on a scale from 0 (little importance) to + + + (very great importance).

Several normal extracellular materials, such as circulating blood components, undergo modifications which accelerate their endocytosis allowing lysosomal digestion. Into this category fall serum glycoproteins: once their terminal sialic acid is removed to expose galactose termini, rapid endocytosis of the degraded glycoprotein molecules by liver hepatocytes occurs. There are also receptors for other (more internal) residues on the carbohydrate side chains, both in liver and other organs. More gross changes, such as complete loss of native structure, may encourage endocytosis. Proteins may also become complexed in the circulation, and several such complexes are particularly susceptible to endocytosis by various cells. Examples are proteinases, which complex with macromolecular inhibitors, and free haemoglobin, which complexes with another serum protein, haptoglobin. Thus abnormal extracellular components arising within the organism are removed and degraded well. In addition, foreign molecules (arising from outside the organism) are also removed and degraded (or sequestered: section 5.2) efficiently, just as are foreign organisms.

Another endocytosed substrate which lysosomes degrade is the plasma membrane. Endocytosis and digestion is continuous in most cells, and as noted earlier, the composition of the endocytosed membrane may not be identical with that of the plasma membrane. This selectivity can allow plasma membrane proteins to have various distinct half-lives. While this is true of plasma membranes in some cell culture systems, there are others in which all the plasma membrane proteins share a single half-life. In these latter cases, endocytosis of membrane with the same composition as plasma membrane occurs, and all the components are presumably degraded as a unit.

The induction of secretion, for instance of pancreatic enzymes, is often accompanied by an increased rate of endocytosis by the cells concerned, which prevents the plasma membrane expanding massively. Much of this endocytosed membrane is degraded by lysosomes, though some may be re-used in the formation of further secretory vacuoles.

The evidence that lysosomes are involved in the degradation of such endocytosed materials is extremely strong. Apart from the quite incriminating evidence that the endocytosed material does enter lysosomes, one impressive experiment can be described. This exploits the fact that, as mentioned earlier, if one has a completely pure protein, one may be able to obtain antibodies specifically reacting with that protein from the serum of an animal injected with the protein. Such specific antibodies were obtained against a major lysosomal proteinase, cathepsin D. Several antibody molecules could combine with each cathepsin D molecule, crowding the surface of the enzyme so much that no protein substrate could gain access to the enzyme. Thus the enzyme could be inhibited. When macrophages were allowed to endocytose haemoglobin (a particularly good substrate for cathepsin D), together with such antibodies, considerable inhibition of subsequent degradation of haemoglobin was observed, when compared with the digestion which occurred in the absence of the antibodies. It was shown that the antibodies had no direct effect on endocytosis, and that they became localized in the same lysosomes as the haemoglobin. This constitutes extremely powerful evidence for the role of lysosomes in the digestion of endocytosed materials.

The wide-ranging digestive potential of lysosomes has been mentioned previously; thus low molecular weight products can be produced from most materials presented to the lysosomal system, and diffuse into the cytoplasm, contributing to cellular nutrition. Nevertheless, in some circumstances lysosomes can be so overloaded with substrates that degradation is insufficiently fast to deal with them. Enlargement and breakage can then occur.

6.2 Degradation of intracellular materials

Lysosomes are also a major agent of degradation of intracellular proteins. Direct evidence for this was obtained by an experiment quite similar to that just described. Whereas antibodies were used in the macrophage system to inhibit cathepsin D, a specific oligopeptide inhibitor of the same enzyme was used in these experiments. It was presented to a perfused rat liver (i.e. a liver isolated from the animal, but with an oxygen and nutrient delivering medium pumping through its blood vessels) in such a way that it was endocytosed rapidly. Having reached the lysosomes it produced a substantial degree of inhibition of the ongoing degradation of intracellular proteins of the liver. Other less direct evidence also supports the notion that lysosomes are important in degradation of intracellular proteins. Whereas lysosomes are the only agent concerned in degrading endocytosed proteins, there are probably also extralysosomal mechanisms for the degradation of intracellular protein, and particularly for the early stages of such degradation.

One of the main characteristics of degradation of intracellular proteins, such as cytoplasmic proteins, is that there is a wide range of half-lives. Lysosomal uptake of soluble cytoplasmic proteins seems superficially to be a random process (Fig. 5–1). But because the proteins may bind to the membrane forming the autophagic vacuole to varying degrees, similarly varied fractional rates (ratio of amount taken up, or degraded, to amount present of uptake, and therefore of degradation, may actually ensue (cf. Fig. 4–3). This selectivity of uptake would result in varied half-lives for the proteins, whereas random uptake by lysosomes would produce a single half-life for all the proteins concerned. Thus lysosomes could possibly be solely responsible for the variation in half-lives of cytoplasmic proteins.

How great a contribution autophagic uptake of intact organelles makes to their turnover (again, their components often show very diverse turnover rates) is unclear, and will remain so at least until a reliable measure of autophagy is available. Because the proteins of most organelles show varied half-lives, degradation cannot be entirely by autophagy of intact organelles; most membrane-bound organelles are thus likely to have some proteolytic system of their own to deal with those proteins which cannot exchange with cytoplasmic pools and thereby gain access to any more selective uptake system. The crinophagic process for dealing with excess secretory vesicles is probably normally a minor component of their turnover. The well-known, though little understood, energy dependence of protein turnover, may be due to the requirement of autophagy for energy.

The important function of lysosomes in regression and remodelling of tissues has been mentioned, and many examples are described in detail in the series of books edited by DINGLE and others. These functions require that lysosomes deal with many substrates besides proteins, and of course they are well equipped to do so. However, there is little information on *in vivo* catabolism of most other materials in normal lysosomes. But several genetic diseases, involving deficiencies of single enzymes, give indications

§ 6.3   EFFECTS OF ACCUMULATION OF MATERIALS BY LYSOSOMES   47

of the range of substrates normally catabolized by lysosomes. Deficiencies of single lysosomal glycosidases, sulphatases, phosphatases, and lipases are known, and the pathological accumulation of materials (glycolipids, gangliosides, mucopolysaccharides, lipids, glycogen) not usually present in large amounts in lysosomes reveals the importance of the enzymes in their normal catabolism (see Fig. 6–2). Many of these substrates arise intracellularly, as they are normal intracellular components; though apparently others arrive by endocytosis. In addition, such syndromes ('storage diseases') have allowed the realization that there are many enzymes detectable only by means of natural substrates, and that synthetic substrates can often be attacked by several distinct enzymes.

A modification of the process of crinophagy in the thyroid gland has been described. Here, endocytosed thyroglobulin undergoes conversion to the normal hormone product in vesicles which receive lysosomal enzymes by fusion. After such conversion, and before complete degradation ensues, the hormone is secreted (Fig. 6–3). This process may also occur in other systems in which intracellular proteolytic activation of protein precursors occurs (proinsulin to insulin (Fig. 6–4), and now many other examples are known) and perhaps in processing of lung mucous material before secretion, activities besides proteolysis may be involved.

6.3   Effects of accumulation of materials by lysosomes

The labilization of lysosomes following accumulation of certain particles (silica, etc.), metals and membrane reactive agents (e.g. retinol) has been mentioned. In several cases subsequent leakage of lysosomal enzymes leads to cell death. The storage of non-digestible particles (e.g. latex) may also be accompanied by secretion of neutral proteinases from specialized granules in some cells.

An extension of these ideas is embodied in the theory of ALLISON concerning carcinogenesis; several carcinogens and transforming viruses accumulate in lysosomes, and slow release of enzymes might cause transformation through the activity of proteinases and nucleases on molecules in the nucleus or cell surface. Activation of lymphocytes (stimulation of DNA synthesis and cell division) and liver regeneration (after removal of part of the liver) both involve perinuclear clustering of lysosomes and thus possibly action of lysosomal enzymes on nuclear material, though this has not been substantiated. By extension, lysosomes may be involved in normal cell division. Release of material from lysosomes can certainly sometimes be compatible with maintenance of cell viability, because several viruses are uncoated in lysosomes, prior to their replication outside lysosomes in viable cells.

Furthermore, SZEGO has shown that target-specific hormones, after endocytosis in target tissues, join the lysosomal system by fusion of the endocytic vesicle with primary lysosomes. The lysosomes, which contains

Disorder	Enzyme deficiency	Metabolites primarily affected
Mucopolysaccharidoses		
Hurler and Scheie syndromes	α-Iduronidase	Proteoglycan components
Hunter syndrome	Iduronate sulphatase	
Sanfilippo syndrome		
A subtype	Heparan N-sulphatase	
B subtype	N-Acetyl-α-glucosaminidase	
Sphingolipidoses		
GM_1 gangliosidosis	β-Galactosidase	GM_1 ganglioside, fragments from glycoproteins and certain complex lipids
Krabbe's disease	β-Galactosidase	
Lactosylceramidosis	β-Galactosidase	
Tay-Sachs disease	Hexosaminidase A	
Sandhoff's disease	Hexosaminidases A and B	
Disorders of Glycoprotein Metabolism		
Fucosidosis	α-L-Fucosidase	Fragments from glycoproteins
Mannosidosis	α-Mannosidase	
Aspartylglycosaminuria	Amidase	
Other Disorders with Single Enzyme Defect		
Pompe's disease	α-Glucosidase	Glycogen
Wolman's disease	Acid lipase	Cholesterol esters, triglyceride
Acid phosphatase deficiency	Acid phosphatase	Phosphate esters
Multiple Enzyme Deficiencies		
Multiple sulphatase deficiency	Sulphatases (arylsulphatase A, B, C: steroid sulphatases; iduronate sulphatase; heparan N-sulphatase)	Steroid sulphatases; proteoglycans
I cell disease and pseudo-Hurler polydystrophy	Almost all lysosomal enzymes except proteinases deficient in cultured fibroblasts; present extracellularly	Proteoglycans and complex lipids
Disorders of Unknown Origin		
Cystinosis	Accumulation of cystine in lysosomes	Cystine
Mucolipidoses I, IV	Ultrastructural evidence of lysosomal storage	Unknown

Fig. 6–2 Selected lysosomal storage diseases and their characteristics.

§ 6.3   EFFECTS OF ACCUMULATION OF MATERIALS BY LYSOSOMES   49

high affinity receptor proteins, then transports the hormone to the nucleus, apparently without degradation. The nuclear membrane invaginates, and lysosomes are internalized; possibly allowing intracellular action of the hormone and/or lysosomal enzymes. This is a verification of some parts of Allison's proposal, although there is still no direct evidence for intranuclear effects of lysosomal enzymes. Open questions in the theory concern the mode of endocytosis of the hormone, i.e. whether a plasma membrane receptor is involved. If so, does this explain the presence of the soluble lysosomal receptor? And how important is this route quantitatively?

DE DUVE has discussed the possible exploitation of the intralysosomal accumulation of drugs, enzymes, inhibitors, metal chelators, liposomes and other agents in therapy of lysosomal storage diseases, cancer and other diseases. For instance, in storage diseases one might wish to replace the missing enzyme in lysosomes of relevant cells. In cases of hyperactivity of individual lysosomal enzymes one might wish to supply a specific inhibitor to the relevant lysosomes to reduce the activity of the enzyme. In

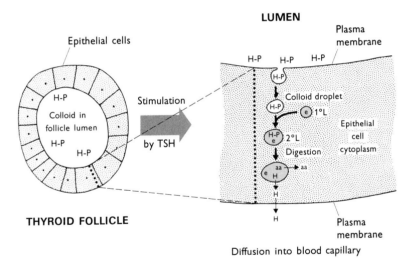

Fig. 6–3 Lysosomal processing in secretion of thyroid hormone. The hormone (H) is initially released into the thyroid lumen, in colloidal form as a complex with a large protein (H–P; thyroglobulin). Following stimulation by thyroid stimulating hormone (TSH), accelerated endocytosis of H–P occurs. The endocytic vesicle fused with primary lysosomes, and degradation of the protein moiety releases free amino acids, and also the low molecular weight hormone (thyroxin). The hormone can then diffuse into the nearest blood capillary, and enter the circulation. *Abbreviations: Within membranes:* aa, amino acids released by digestion; e, lysosomal enzymes. *Other membranes:* 1°L, primary lysosome; 2°L., secondary lysosome.

the case of deleterious accumulation of metals in lysosomes, one might wish to supply agents which will bind the metals specifically (chelators), and thus prevent their undesirable intralysosomal effects. And in cancer one would hope to be able to direct agents capable of killing cells specifically to the cancer cells.

Many of the potentially useful agents (such as the replacement enzymes) will anyway accumulate in lysosomes by the mechanisms discussed already. Others, if not capable of penetrating membranes, may be directed there by means of liposomes, while penetrating agents may be complexed to larger, non-penetrating, agents which will carry them to

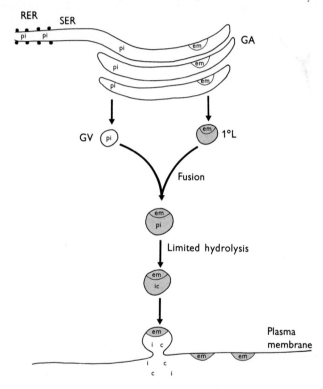

Fig. 6–4 A possible mechanism of insulin biosynthesis by pancreatic β-cells. Proinsulin (pi) is formed on rough endoplasmic reticulum (RER) and transferred via smooth endoplasmic reticulum (SER) and the Golgi apparatus (GA) to Golgi vesicles (GV). These fuse with primary lysosomes (1°L) containing some membrane-bound proteinases (em) which cleave the proinsulin molecule at a limited number of sites to release a 'connecting peptide' (c) and native insulin (i). Both products are released by exocytosis, while the proteinases may remain bound to the plasma membrane, and subsequently be re-endocytosed. *Abbreviations: Within membranes:* em, membrane-bound lysosomal enzyme.

lysosomes, or together with which they may be supplied in liposomes. Other aspects of lysosomal behaviour, such as membrane fusion, may be accessible to control by means of such approaches.

Clearly many such materials can enter cells of several types, and as tumours have high pinocytic rates they may form a good target. Means of ensuring specificity of uptake by the desired target cells by incorporating antibodies or other biological macromolecules with specificity for components of the target cell plasma membrane into the carrier molecule or liposome are being sought. Drugs can often be liberated in active form after digestion of the carrier within the lysosome. However, the blood-brain barrier (a permeability barrier to substances of high molecular weight) remains a considerable problem in dealing with those many diseases in which lesions to brain tissue, as well as to others, occur.

6.4 Extracellular activities

Exocytosed lysosomal enzymes may cause a variety of degradative processes. Thus, the first stage of connective tissue breakdown, both in remodelling prior to ossification and in arthritic conditions, occurs extra-cellularly and lysosomal enzymes are probably involved. Extracellular lysosomal enzymes produced by tumour cells may facilitate their invasion of tissues.

Lysosomal enzymes are thought to be important mediators of inflammation after release stimulated by immune complexes adhering to basement membranes, cartilage, etc. Some neutral proteinases are important, but acid proteinases may also contribute. At present, it is difficult to evaluate the relative importance of individual proteinases, though some can produce inflammatory lesion when injected alone; and local pH values are unknown. Diverse connective tissue components (basement membranes, cartilage, arterial walls, etc.) are attacked, but also cascade processes are induced by lysosomal enzymes: the blood proteinase cascade ('complement'), end products of which can lyse cells, can be activated and give rise additionally to factors which both attract leucocytes and induce lysosomal exocytosis; in both ways, the system encourages further lysosomal enzyme release.

The oligopeptide bradykinin, which dilates some blood vessels and induces increased vascular permeability allowing an outflow of fluids, is generated from a precursor by proteolytic cleavage, and some lysosomal proteinases are capable of this conversion. Several other pharmacologically active peptides may perhaps be generated as well as destroyed by extracellular lysosomal enzyme activity. Furthermore, a cationic protein from neutrophil lysosomes is a potent releaser of histamine from mast cells, and thus contributes further to the elevation of vascular permeability.

Some extracellular proenzymes, which are inactive until they have undergone restricted proteolytic cleavages, may be activated by secreted lysosomal proteinases. On the other hand, recent evidence suggests that the macrophage activator of the precursor (plasminogen) of the fibrinclot digesting enzyme, plasmin, is non-lysosomal, although an activator known in several tissues may be lysosomal. Released kidney lysosomal enzymes may be concerned in the generation of erythropoietin, the stimulator of erythropoiesis, the process of formation of erythrocytes. But the opposed process of haemolysis can be performed by lysosomal enzymes, and viral haemolysis may be due to lysosomal enzymes attached to the virus particles during intracellular development.

Non-immunologic killing of tumour cells by macrophages apparently proceeds by direct secretion of lysosomal enzymes into the target cell: membrane stabilizers inhibit, but the initiating factors, presumably surface characteristics of the tumour, are unknown. This injection mechanism may be involved in some types of immunospecific cell killing by lymphocytes. Plants also exploit lysosomal enzymes in their defences against invasion: defence against pathogenic fungi often involves the induction of lysosomal enzyme secretion by the host.

Related surface factors are probably (1) involved in controlling intercellular adhesion before fusion, and (2) responsible for differences in growth control between cells which have been briefly subject to proteolysis, without being killed, and cancer cells on the one hand, and normal cells on the other. In the case of cell fusion it has been shown that surface modifications increasing fusion can be produced by released lysosomal enzymes, and similar lysosomal mechanisms may induce lymphocyte and other cell transformations and detachment of tumour cells prior to metastatic spread.

Glossary of Lysosomal Terminology

Autophagic Vacuoles: Membrane lined vacuoles containing morphologically recognizable cytoplasmic components. Comprise autolysosomes (which are secondary lysosomes, q.v.) and autophagosomes (which are vesicles sequestering cytoplasmic organelles which have not yet received lysosomal enzymes). Synonyms for autophagic vacuoles include cytolysome and cytosegresome.

Autophagy: The process of sequestration of intracellular components in vacuoles (including autophagic vacuoles, q.v.) which become lysosomes.

Crinophagy: A specialized form of autophagy in which secretory vesicles, normally carrying a cellular product to the exterior of the cell, instead fuse with lysosomes and are degraded.

Endocytosis: Internalization of formerly extracellular material, within a membrane-bound vesicle formed by invagination of the plasma membrane.

Exocytosis: Release of vesicular contents to the extracellular medium, by fusion of the vesicle membrane with the plasma membrane.

Heterolysosomes: Secondary lysosomes (q.v.) containing substrates derived from outside the cell by endocytosis (see Chapter 4), formed by fusion of primary lysosomes with heterophagosomes carrying the substrates.

Multivesicular body: Autophagic vacuole lined by a single membrane and containing inner vesicles resembling Golgi vesicles. They contain lysosomal enzymes, and so are a form of lysosome. They are formed by invagination of the external membrane, which then buds inwards to form an autophagic vacuole containing lysosomal membrane and some cytoplasmic material (see Fig. 5–1).

Phagocytosis: A form of endocytosis in which particulate material is taken up into large vesicles.

Pinocytosis: Endocytosis of soluble materials into small vesicles.

Primary Lysosomes: Lysosomes containing active acid hydrolases, which have not yet undergone fusion with other vesicles to bring them into contact with substrates.

Residual Bodies: Secondary lysosomes (q.v.) containing undigested residues (membrane fragments and whorls).

Secondary Lysosomes: The product of fusion of a primary lysosome with other intracellular vesicles containing substrates. Examples are heterolysosomes (q.v.) and autolysosomes.

Secretion: Release of cellular products into the extracellular space. Exocytosis is a specialized form.

Further Reading

Introductory Material
LOCKWOOD, A. P. (1971). *The Membranes of Animal Cells.* Studies in Biology 27. Edward Arnold, London.
The Cell, Readings from *Scientific American.* Freeman, San Francisco.

Broad Reviews
DE DUVE, C. and WATTIAUX, R. (1966). *Ann. Rev. Physiol.*, **28**, 435–492. The classic general review.
ALLISON, A. C. (1968). *Advan. Chemother.*, **3**, 253–302. Concerning drugs and lysosomes.
WILSON, C. L. (1973). *Ann. Rev. Phytopathol.*, **11**, 247–272. Plant lysosomes.
DE DUVE, C., DE BARS, T., POOLE, B., TOUET, A., TULKENS, P. and VAN HOOF, F. (1974). *Biochem. Pharmacol.*, **23**, 2495–2531. Mechanisms of accumulation by lysosomes.
NEUFELD, E. F., LIM, T. W. and SHAPIRO, L. J. (1975). *Ann. Rev. Biochem.*, **44**, 357–376. Lysosomal storage diseases.
LOCKSHIN, R. and BEAULATON, A. (1974). *Life Sciences*, **15**, 1549–1565. Lysosomes and cell death in normal and pathological processes.

Books
DINGLE, J. T. and FELL, H. B. (eds.) (1969a). *Lysosomes in Biology and Pathology, Vol. 1*. North Holland, Amsterdam.
DINGLE, J. T. and FELL, H. B. (eds.) (1969b). *Lysosomes in Biology and Pathology, Vol. 2*. North Holland, Amsterdam.
DINGLE, J. T. (ed.) (1973). *Lysosomes in Biology and Pathology, Vol. 3*. North Holland, Amsterdam.
DINGLE, J. T. and DEAN, R. T. (eds.) (1975). *Lysosomes in Biology and Pathology, Vol. 4*. North Holland, Amsterdam.
DINGLE, J. T. and DEAN, R. T. (eds.) (1976). *Lysosomes in Biology and Pathology, Vol. 5*. North Holland, Amsterdam.
DINGLE, J. T. (ed.). *Lysosomes, a Laboratory Handbook.* (2nd ed.) North Holland, Amsterdam, in press. This series of books provides detailed reviews on nearly every aspect of lysosomal physiology and pathology.
HERS, G. and VAN HOOF, F. (eds.) (1973). *Lysosomes and Storage Diseases.* Academic Press, New York.
HOLTZMANN, E. (1976). *Lysosomes: a Survey.* Springer-Verlag, Vienna.